Fluorescent ANA's and Beyond!

by

Carol Peebles, MS, MT (ASCP)

FLUORESCENT ANA'S AND BEYOND!
by Carol Peebles, MS, MT (ASCP)

ISBN: 1-60039-088-9

INOVA Diagnostics, Inc.
9900 Old Grove Road
San Diego CA 92131-1638 USA
Phone: 858.586.9900 Fax: 858.586.9911
www.inovadx.com

690002 February 2009 Rev. 0

About the Author

Carol Peebles received her BA and MS from Kansas State College of Pittsburg in Pittsburg, KS. She received her Medical Technology certification from St. Francis Hospital in Wichita, KS. In 1974 she became supervisor in the laboratory of Eng M. Tan, an expert in autoimmune disease. In 2001 she joined the staff of INOVA Diagnostics, Inc. She is co-author on numerous publications.

CONTENTS

CHAPTER 3

GLOSSARY

ACA	-	anti-centromere antibodies
AF	-	Arthritis Foundation
ANA	-	anti-nuclear antibodies
ANCA	-	anti-neutrophil cytoplasmic antibodies
CDC	-	Centers for Disease Control
CENP	-	centromere protein
CIE	-	countercurrent immunoeletrophoresis
CREST	-	syndrome of Calcinosis, Raynaud's phenomenon, Esophageal dysmotility, Sclerodactyly and Telangiectasias
DABCO	-	1,4-diazobicyclo-(2,2,2)-octane. Flourescent anti-fading agent
DNA	-	deoxyriboneucleic acid
dsDNA	-	double stranded DNA
EIA	-	enzyme immunoassay
EJ	-	glycyl-tRNA synthetase
ELISA	-	enzyme linked immunosorbent assay
ENA	-	extractable nuclear antigen
ER	-	endoplasmic reticulum
FITC	-	flourescein isothiocyanate
HEp-2	-	human epithelial cell line: type 2
HEp-2000	-	transgenic HEp-2 cells expressing increased SSA
hnRNP	-	heterogenous ribonuclear proteins
IFA	-	immunoflourescence assay
Ig	-	immunoglobulin
IIF	-	indirect immunoflourescence
IUIS	-	International Union of Immunological Societies
Jo-l	-	ENA, histidyl-tRNA synthetase
kDa	-	molecular weight in kilo-Daltons
La	-	index patient with anti-SSB antibody
LAMPs	-	lysosomal-associated membrane proteins
LAP	-	lamin associated protein
L.E. cell	-	lupus erythematosus cell
MCTD	-	mixed connective tissue disease
MSA	-	mitotic spindle apparatus
NOR-90	-	nucleolar organising region protein 90kDa
NuMA	-	nuclear mitotic spindle apparatus
OJ	-	isoleucyl-tRNA synthetase
PBC	-	primary biliary cirrhosis
PBS	-	phosphate buffered saline
RBC	-	red blood cell

Chapter I

Evaluation and Interpretation of Indirect Immunofluorescence on HEp-2 cell Substrates

Evaluation and Interpretation of Indirect Immunofluorescence on HEp-2 cell Substrates

In 1943, Dr. Malcolm Hargraves recorded an observation of "peculiar rather structureless globular bodies taking purple stain (artifact) in a bone marrow report on a 9 y/o child with an obscure clinical problem." [1] This was the first observation of what was to be identified as the LE cell phenomenon. In 1948, he published the first description of the LE cell. He went on to demonstrate that the phenomenon was not limited to the bone marrow but could be produced using sera from patients diagnosed with systemic lupus erythematosus (SLE) [2]. Further research demonstrated that the factor responsible for inducing the phenomena was IgG in the presence of adequate amounts of complement. In 1957, three different groups of researchers reported the application of the indirect immunofluorescent method for the detection of bound immunoglobulins that was developed by Coons et. al. [3,4] to the study of sera from SLE patients [5,6,7]. The method gained clinical relevance in the mid to late 1960's [8,9,10]. (Figure 1)

The substrates in common use at this time were primarily rodent tissues, mouse kidney or rat liver. Others included spleen touch preps, chick cell nucleated RBC and occasionally tissue culture cells. Parker and Kerby, in 1974, described the use of monolayer tissue culture cells in the evaluation of connective tissue diseases [11]. They reported that ANA patterns on the cells were sharper than on cryostat- sectioned tissues. While they tested for both IgG and IgM immunoglobulins, they chose to report only the IgG results as they found "IgM ANA titres and patterns were of limited clinical value." Soon after, several companies began to market ANA test kits utilizing tissue culture cells as a substrate. Today, the substrate of choice for ANA detection is HEp-2 cells.

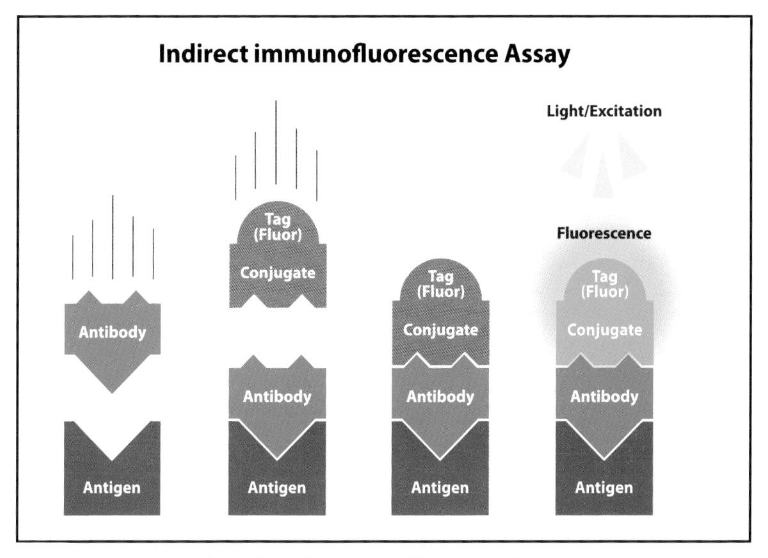

Figure 1. Indirect immunofluorescence Assay.

There are several advantages of HEp-2 cells over tissue sections. First, they are individual cells grown and fixed on a flat surface making the antigens more uniform from cell to cell. They are actively growing cells with a large nucleus making them easier to read. Also since they are actively replicating they contain antigens not present in tissue sections. The specific identification of centromere patterns as well as certain cell cycle patterns would be impossible without the use of HEp-2 cells as a substrate.

While there are many different HEp-2 cell kits on the market, they do not all give comparable results. Standardization between laboratories is difficult. Major reasons for the differences include the growth conditions, fixation and preservation of the HEp-2 cells, the characteristics of the immunoglobulin conjugates included in the kits as well as the optical systems used by individual laboratories. A brief discussion of these variables will follow.

Cells should be grown in such a manner that they include adequate dividing cells for proper evaluation and identification of specific patterns including centromere and various cell cycle patterns.

Cells should be fixed in such a way as to preserve important antigens. The common fixatives for the cells include ethanol or methanol, acetone, and combinations of these for different times and at different temperatures. As an example of the effect of fixation on specific antigens, it has been demonstrated that the SS-A/Ro antigen is affected by prolonged alcohol fixation [12]. Fixation in alcohol decreases or eliminates its detection (Figure 2). Since antibodies to SS-A/Ro are important in SLE and Sjögren's syndrome (SS) and may be the only antibodies detected in subacute cutaneous LE and neonatal lupus, it is

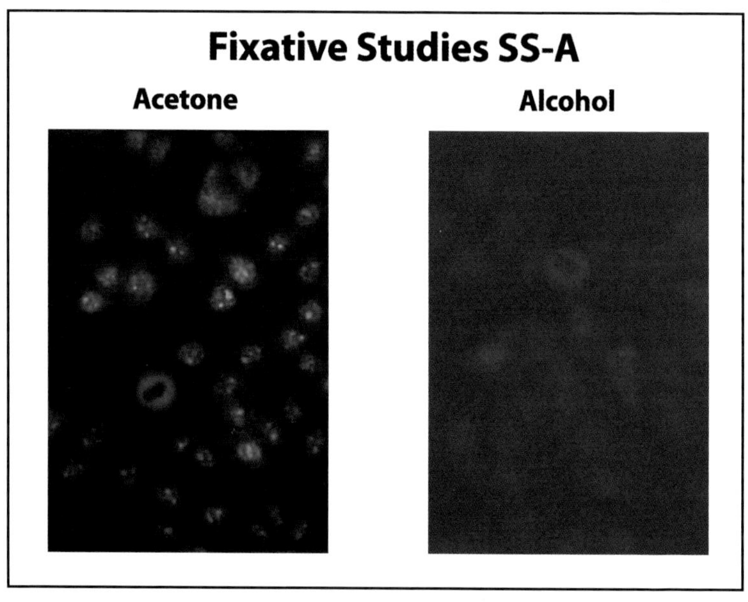

Figure 2. Fixation comparison of SS-A/Ro on HEp-2 cells.

important to detect them. To verify if the cells in use detect antibodies to SS-A/Ro, the inclusion of a serum positive for these antibodies is recommended as a standard control with each test procedure.

While substrates vary considerably, it is the variation in conjugates that appears to be mainly responsible for inconsistent results in IIF testing according to the CDC [13 p.35]. Two major variables leading to these differences include the specificity and the sensitivity of the conjugates used for detection. The conjugates used in kits today are of two major types. The first are class specific conjugates reacting only with a specific immunoglobulin class, i.e. IgG, IgA or IgM. These are labeled with the immunoglobulin class as Fc specific. The reactivity of these antibodies is directed to epitopes on the constant region of the heavy chains (See Figure 3). The second type of conjugate is one that reacts with more than one class of immunoglobulins. The methods used in the preparation will yield variable results in the balance of antibodies to the different immunoglobulin classes. The most commonly used conjugate in the early studies of ANAs was to the whole IgG molecule using Cohn

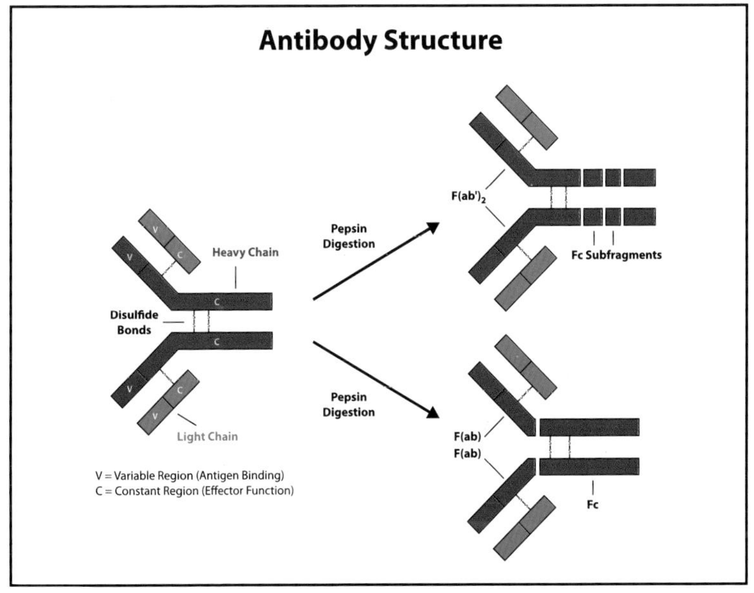

Figure 3. Antibody structure.

fraction II as the immunogen. These conjugates react not only with the IgG heavy chain but also with IgA and IgM due to the presence of antibodies to epitopes on the light chain or F (ab) regions common to all immunoglobulin classes. These are labeled as antibodies to IgG (H&L). While they detect more than one class of immunoglobulin, the amount of light chain reactivity is highly variable. A better method for preparing a conjugate to total immunoglobulins is by conjugation of antibodies elicited in animals immunized with total immunoglobulins. These conjugates include antibodies that are directed to the various heavy chain epitopes of the immunoglobulin classes and do not rely solely on the light chain cross reactivity. A third and perhaps the best method of preparing conjugates reacting with all classes of immunoglobulins are by mixing individual class specific conjugates together in a balanced fashion. There are differences of opinion as to which of these is best. While the CDC handbook from 1987 suggests the use of a conjugate directed to total immunoglobulins, current authors recommend the use of IgG specific conjugates. Hollingsworth et. al. [14] in their chapter on antinuclear antibody testing advocate the use of IgG specific conjugates but state if it is necessary to detect IgA and IgM, an isotype-specific blend of IgG, IgA and IgM should be used. Humbel [15], in 1993, in the *Manual of Biological Markers of Disease* states that antihuman IgG Fc-specific is now the preferred fluorochrome conjugate. The same recommendation is made by Kavanaugh et.al [16] in the *Guidelines for Clinical use of the Antinuclear Antibody Test and tests for Specific Autoantibodies to Nuclear Antigens.* The NCCLS guidelines [17] states, "With use of an anti-IgG specific conjugate, the IgM-class antinuclear antibodies associated with rheumatoid arthritis, drugs, and age, which are usually of no diagnostic significance, are not detected." The experience in the Duke University FANA laboratory suggests the use of a monospecific anti-IgG Fc specific conjugate may lead to increased specificity, since most ANA pathological significance are of the IgG class [17, p.60]

The detection of antibodies specific for IgA or IgM without having antibodies to IgG is rare [19,20,21]. As mentioned previously, Parker and Kerby [11] testing for IgG and IgM antibodies determined that aside from some shaggy IgM staining with certain RA sera, the determination of IgM ANA yielded no clinically useful information. An exception to IgG antibodies being present along with IgM antibodies was reported in 1986 by Molden et. al. These antibodies were relatively rare and occurred in patients with undifferentiated connective disease. They were unusual as they stained primarily the nuclei of differentiated cells in tissue sections as well as the connective tissue nuclei in the spleen and lymph nodes in a variable large speckled pattern but did not stain the nuclei of lymphocytes. These antibodies are not detected in HEp-2 cell nuclei on conventional ANA slides.

It is known that in the primary immune response IgM antibodies develop first, with a subsequent isotype switch to IgG as the response continues. The secondary response is an IgG response with increased specificity. In the systemic rheumatic diseases the antibody response is compatible with an antigen driven response resulting in the production of IgG antibodies. It can therefore be noted that while the use of IgG specific conjugates may miss a few patients, those with clinically relevant antibodies will be detected

Another major variable that accounts for differences in test results, especially the titer, is the number of fluorescein molecules per antibody molecule (F/P ratio). Obviously, the greater the F/P ratio is the more sensitive the assay. While this can be advantageous, it must be balanced by the fact that the greater the F/P ratio the greater the chance for nonspecific staining. The F/P ratio optimum recommended by NCCLS [17] is 3. The CDC Handbook [13] states 3-5. It should be noted that conjugates supplied by the manufacturers are optimized for their substrates and may not yield equivalent results on other manufacturers substrates. If employing a conjugate not directly supplied by the manufacturer, the laboratory must perform a chessboard titration and optimize the dilution of the conjugate to each individual assay in which the conjugate is utilized. As should be apparent from the above examples, one of the most important questions to ask in evaluating discrepancies between laboratories on FANA results is "what is the specificity and F/P ratio of the conjugate that each laboratory uses?"

In order to assure comparable results in day-to-day testing, the addition of a specified volume of the diluted patient serum/well is important. The limiting factor in the reaction is the amount of specific antibody in the patient's sera. Additionally, the amount of conjugate added should be sufficient to completely cover the slide. The screening dilutions and normal limits MUST be established in each laboratory.

A suggestion for titering positive sera to obtain relevant information without using excess wells is to test using 4-fold dilutions, i.e. 1:40, 1:160 and 1:640. This gives a value in the low, the mid and the high range. Reporting the patterns and relative intensity at each dilution provides ample information for the physician. In a study to determine the range of ANAs in healthy individuals involving 15 international laboratories performing ANA studies on HEp-2 cells, Tan et. al. [23] tested samples that included sera from healthy individuals in 4 age subgroups spanning 20-60 years of age and patients with SLE, SSc, SS, RA or soft tissue rheumatism. In this putatively normal population, the ANA was positive 31.7% at 1:40. 13.3% at 1:80, 5% at 1:160, and 3.3% at 1:320. The 1:40 dilution while detecting NHS at 31.7% also detected virtually all of the SLE, SSc and SS as ANA positive yielding a high sensitivity, but low specificity. If a cutoff at a 1:160 serum dilution was used, only a portion of the disease patients would be detected but 95% of the NHS would be excluded yielding a high specificity and lower sensitivity. They therefore recommended that "ANA tests should report results at both the 1:40 and 1:160 dilutions, and should supply information on the percentage of normal individuals who are positive at these dilutions."

As mentioned before, microscopes are highly variable from laboratory to laboratory. In general, they are either incident or transmitted light and use either halogen or mercury bulbs. The selection of a microscope depends on the needs and resources of the individual. ANA relative intensity is graded from 1+ (lowest specific fluorescence that enables the nuclei to be clearly differentiated from the background fluorescence) to 4+ (brilliant apple green fluorescence). ANA titer results are usually reported as the highest titer giving a 1+ reaction. For quality control at this step, the laboratory should set exact parameters as to what magnification to use to read titers and patterns and to determine what is considered a 1+ reaction. Standardization between individuals interpretation of the slides is important. Including a titered control with a defined endpoint may help accomplish this.

The interpretation of patterns, while seemingly complex, is simple if one follows certain guidelines. First, examine the entire well. The cells on the slide will have a variable morphology. The majority of the cells will be in the interphase or G1 stage of the cell cycle with the characteristic morphology of a eukaryotic cell (Figure 4). The variable

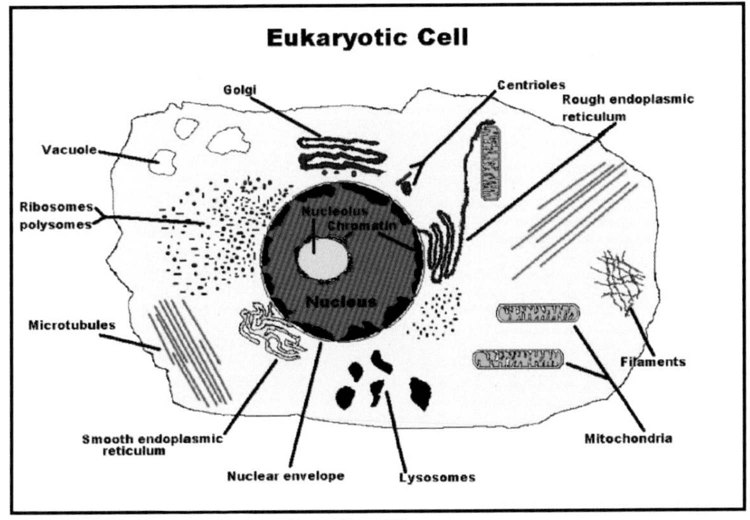

Figure 4. Cell Structure

morphology seen in some cells is due to the cells undergoing cell division. The stages of the cell cycle are diagramed in Figure 5 with Figure 6a-e demonstrating the morphology of cells during mitosis. Second, note the relative intensity of the staining and variation in pattern between the interphase cells. If the staining of the interphase cells is relatively consistent, report the patterns observed. Cells may demonstrate either one pattern (homogeneous, speckled, nucleolar, centromere, etc.) or a mixed pattern with more than one pattern observed (homogeneous and nucleolar, speckled and nucleolar, centromere and mitochondrial, etc.). In some instances, components of the cytoplasm may stain (ribosomes, mitochondria, golgi, spindles, etc.) While antibodies to these cytoplasmic components may

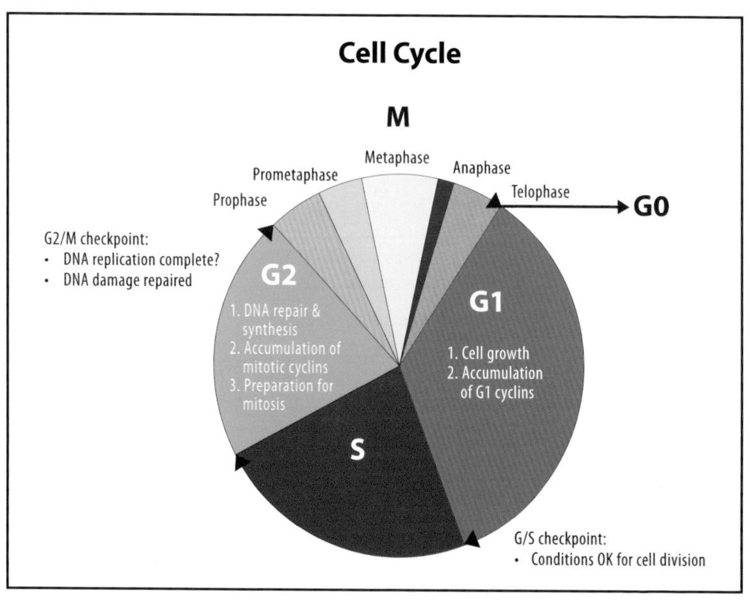

Figure 5. Cell Cycle

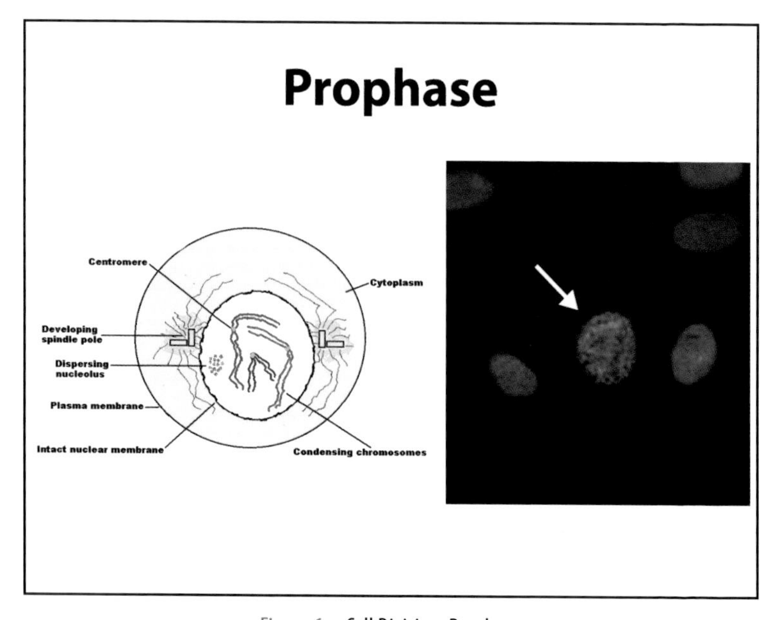

Figure 6a. Cell Division: Prophase

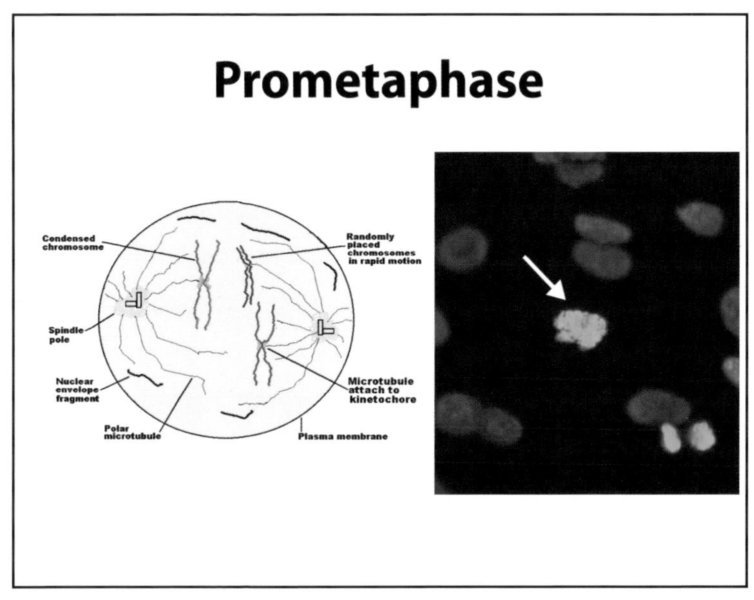

Figure 6b. Cell Division: Prometaphase

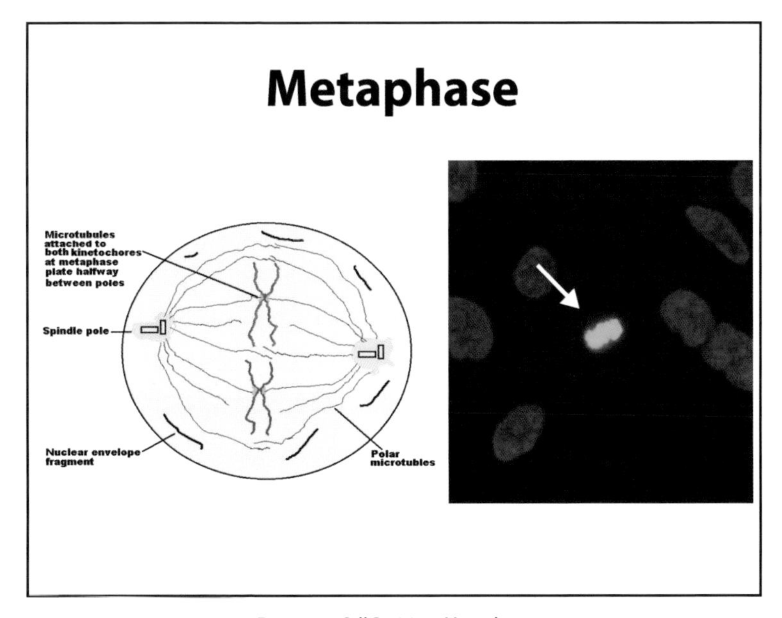

Figure 6c. Cell Division: Metaphase

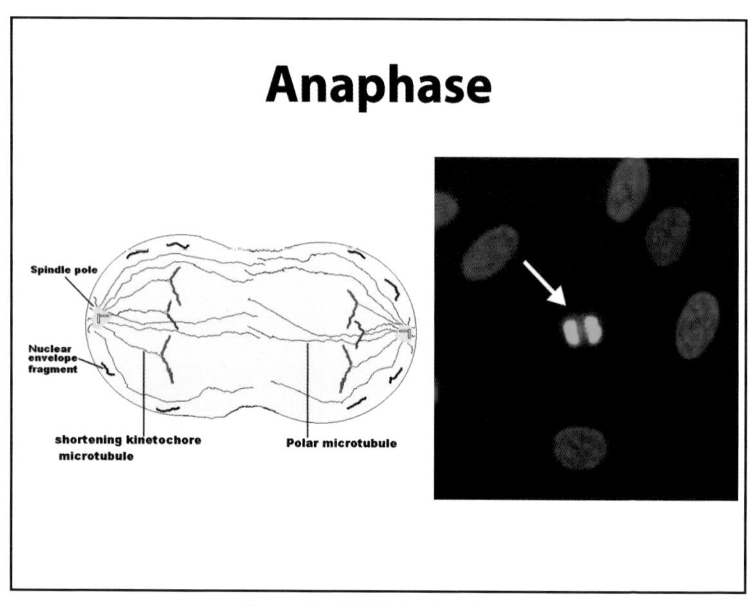

Figure 6d. Cell Division: Anaphase

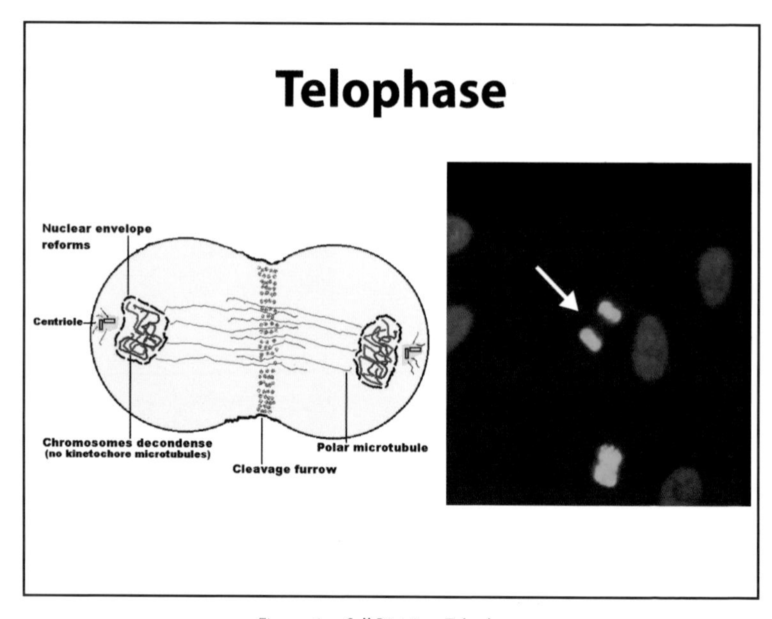

Figure 6e. Cell Division: Telophase

not technically be a part of ANA screening, some do have clinical relevance and should be reported. Examples of patterns are found in the appendix. Third, look at the chromosome area of the mitotic cells. If the reactivity remains on the chromosomes through out the cell cycle it means that while the antibodies may be detecting DNA and /or chromatin, they may also be reacting with one of the many antigens that remain associated with chromatin through out the cell cycle. These latter antigens are present in a lesser concentration than DNA and chromatin and therefore produce a dense fine speckled pattern on both the interphase cells as well as the dividing cells. This is one of the most common patterns observed in the clinical laboratory. The identity of most of these antigens is unknown.

A speckled ANA pattern with little or no staining of the chromosomes during mitosis is characteristic of a group of antibodies that react with extractable nuclear antigens (ENA). The most frequently identified antibodies among these are RNP, Sm, SS-A/Ro and SS-B/La. Antibodies to centromere give a characteristically discrete speckled pattern in the interphase cells with discrete speckles decorating the centromeric region of the chromosomes throughout cell division. By observing both the interphase and metaphase staining of the cell cycle, much information can be obtained on the identity of the antigen being detected.

Marked variations in intensity or pattern between cells are indicative of antibodies reactive with proteins specifically expressed in different stages of the cell cycle. Antibodies to PCNA, the best known of these, are detected in SLE patients. The PCNA antigen is expressed in trace amounts in the G phase of the cell cycle. It is up regulated during the S phase of the cell cycle and, as an accessory factor to DNA polymerase delta, is associated with active sites of DNA replication that leads to the irregular staining and variation intensity observed. Other cell cycle antibodies may be associated less with systemic rheumatic diseases and more with cancer. An example of this is CENP-F (p330[d]), an antibody to a cell cycle centromere related antigen [24]. This antigen accumulates in the nuclear matrix during S phase with the maximum level of intensity detected in the G2/M stages of cell division. It is associated with the centromeres during the early stages of mitosis and with the later stages becomes associated with the spindle midzone and the midbody regions. There are other rarely encountered antibodies to cell cycle proteins. In reporting these it is probably best to report just the presence of a cell cycle antibody unless specific identification is made. Examples of various patterns are included. By becoming familiar with the different patterns of staining and relating them to the results of specific antibody tests (Figure 7), one can better evaluate the overall results of each patient.

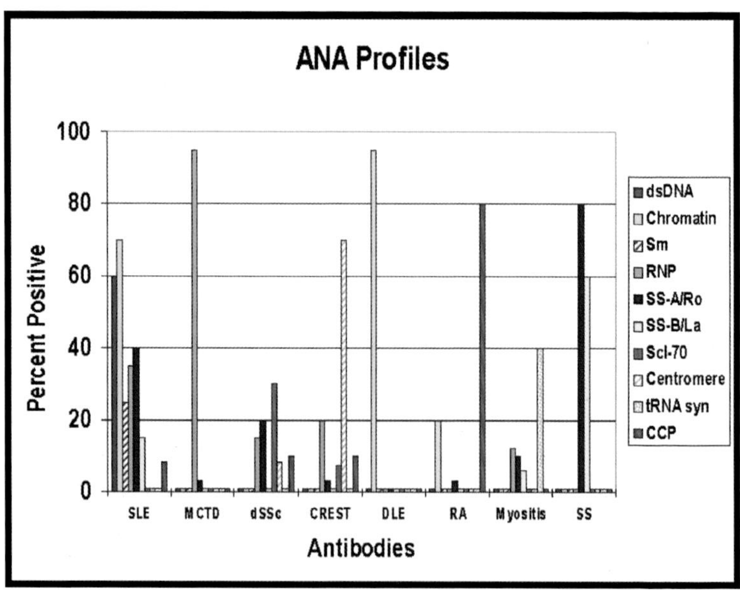

Figure 7. ANA Profiles

Bibliography

1. Hargraves MM, H .Richmond, R Morton. Presentation of Two Bone Marrow Elements: the "Tart" Cell and the "LE" Cell, *Mayo. Clin. Proc.* 23: 25-28. 1948.

2. Hargraves MM. Production In Vitro of the L.E. Phenomenon: Use of normal bone marrow elements and blood plasma from patients with acute disseminated lupus erythematosus, *Mayo Clin Proc* 24: 234-237. 1949.

3. Coons AH, MH Kaplan. Localization of Antigen in tissue Cells: II. Improvements in a method for detection of antigen by means of fluorescent antibody. *J. Exper. Med.* 91: 1-13. 1950.

4. Weller TH and AH Coons. Fluorescent antibody studies with agents of varicella and herpes zoster propagated *in vitro*. *Proc. Soc. Exp. Biol. (N.Y.)* 86: 789-794. 1954.

5. Holman H and HG Kunkel. Affinity between the lupus erythematosus serum factor and cell nuclei and nucleoprotein, *Science* 126:162. 1957.

6. Friou GJ. Clinical application of lupus serum—nucleoprotein reaction using the fluorescent antibody technique, (Abstr.) *Jour Clin Invest* 36: 890. 1957.

7. Holborow EJ, DM Weir, GD Johnson. A serum factor in lupus erythematosus with affinity for tissue nuclei. *Brit. Med. J* 5047: 732-734. 1957.

8. Kunkel HG and EM Tan. Autoantibodies and Disease. *Adv. Immunol* 4: 351- . 1964.

9. Beck JS. Antinuclear antibodies: methods of detection and significance. *Mayo Clinic Proc* 44: 600-619. 1969.

10. Ritchie RF. The clinical significance of titered antinuclear antibodies. *Arth. Rheum.* 10:554-552. 1967.

11. Parker MD, GP Kerby. Combined titre and fluorescent pattern of IgG antinuclear antibodies using cultured cell monolayers in evaluating connective tissue diseases, *Ann. Rheum. Dis* 33: 465-472. 1974.

12. Harmon C, JS Deng, CL Peebles, EM Tan. The importance of tissue substrate in the SS-A/Ro antigen-antibody system. *Arthritis Rheum* 27:166-173. 1984.

13. Cavello JJ. Determination of optimal conjugate dilutions. In *Laboratory Methods for the Detection of Antinuclear Antibodies.* Eds. Cavallaro et.al, US Dept. HHS. 1987. pp33-39

14. Hollingsworth PN, SC Pummer, RL Dawkins. Antinuclear antibodies. In: *Autoantibodies.* Peters JB, Y Schoenfeld eds. Elsiever Science N.Y. 1996. pp74-90

15. Humbel RL. Detection of antinuclear antibodies by immunofluorescence. In: Manual of Biological Markers of Disease. Kluwer Academic Publisher. The Netherlands. 1993. ppA2: 1-16

16. Kavanaugh A, R Tomar, J Reveille, DH Solomon, and HA Homberger. Guidelines for clinical use of the antinuclear antibody test and tests for specific autoantibodies to nuclear antigens. *Arch Path& Lab Med* 124: 71-81. 2000.

17. Quality Assurance for the Indirect Immunofluorescence Test for Autoantibodies to Nuclear Antigens (IF-ANA); Approved Guidelines. NCCLS I/LA2-A Vol. 16, No.11. 1996.

18. McCarty GA, DW Valencia, MJ Fritzler. Antinuclear Antibodies: Contemporary Techniques and Clinical Application to Connective Tissue Diseases, Oxford Press. 1984.

19. Gonzalez EN and NF Rothfield. Immunoglobulin class and pattern of nuclear fluorescence in systemic lupus erythematosus. *NEJM* 274:1333-1338. 1966.

20. Bickel YB, EV Barnett, CM Pearson. Immunofluorescent patterns and specificity of human antinuclear antibodies. *Clin Exp Immunol* 3:641-656. 1968.

21. Aitcheson CT, C Peebles, F Joslin, and EM Tan. Characteristics of antinuclear antibodies in Rheumatoid Arthritis. *Arth. Rheum.* 23:528-538. 1980.

22. Molden DP, GL Klipple, CL Peebles, RL Rubin, RM Nakamura, and EM Tan. IgM anti-histone H-3 antibody associated with undifferentiated rheumatic disease syndromes. *Arth Rheum* 29:39-46. 1986.

23. Tan EM, TEW Feltkamp, JS Smolen, B Butcher, R Dawkins, MJ Fritzler, T Gordon, JA Hardin, JR Kalden, RG Lahita, RN Maini, JS McDougal, NF Rothfield, RJ Smeenk, Y Takasaki, A Wiik, MR Wilson, JA Koziol. Range of Antinuclear Antibodies in "Healthy" Individuals. *Arth Rheum* 40: 1601-1611. 1997.

24. Casiano CA, G Landburg, RL Ochs, and EM Tan. Autoantibodies to a novel cell cycle-related protein that accumulates in the nuclear matrix during S phase and is localized in the kinetochores and spindle midzone during mitosis. *Jour Cell Sci* 106:1045-1056. 1993.

Appendix

HEp-2 Patterns

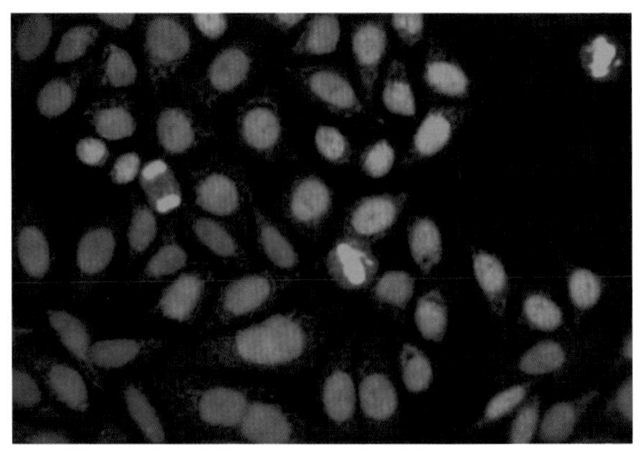

DNA/Chromatin antibodies. The interphase nucleus is homogeneous. The metaphase chromosomes of the dividing cells are also positive. Note the coarse speckling in the cytoplasm. This represents antibodies to mitochondrial DNA. These antibodies are associated with SLE and drug induced lupus.

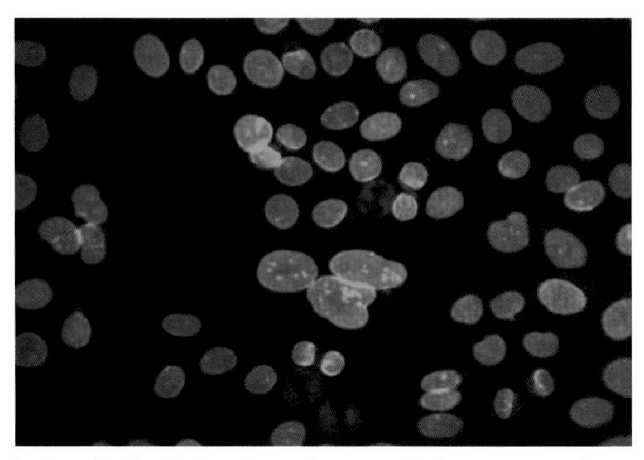

Lamin antibodies. The interphase cells appear to be homogeneous or have a fine ground glass appearance. Note the enhanced rim of some of the nuclei especially where the nuclei are next to each other. The metaphase chromosomes of the dividing cells are unstained. The lamin antibody reacts with an antigen in the nuclear membrane of the cell. These antibodies have been reported in autoimmune hepatitis, primary biliary cirrhosis (PBC) and in patients with antiphospholipid antibodies.

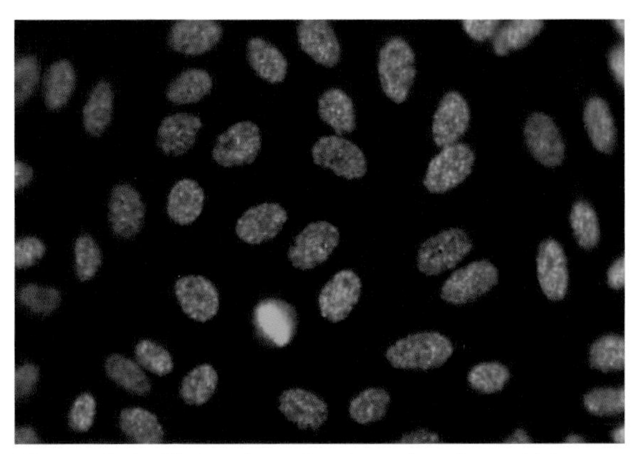

DFS-70 antibodies. The interphase cells stain in a dense fine speckled pattern. The staining of the metaphase cells reflects the dense fine speckled pattern. This pattern is sometimes reported as homogeneous and sometimes speckled. These antibodies represent one of many different antibodies reacting with DNA binding proteins. They have been associated with patients with atopic dermatitis and undifferentiated connective tissue disease.

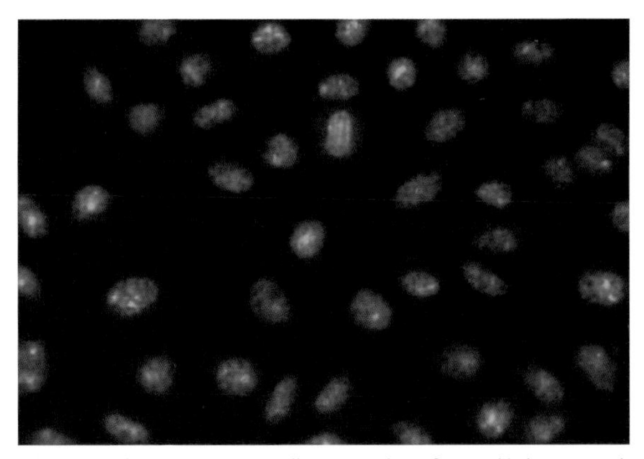

Scl-70 antibodies. The interphase cells stain in a dense fine speckled pattern with many cells having a weak granular nucleolar stain that blends with the nuclear staining. This reaction is serum and fixative dependent. The chromosomes of the dividing cells demonstrate a dense fine speckled pattern. These antibodies are associated with diffuse progressive systemic sclerosis (PSS).

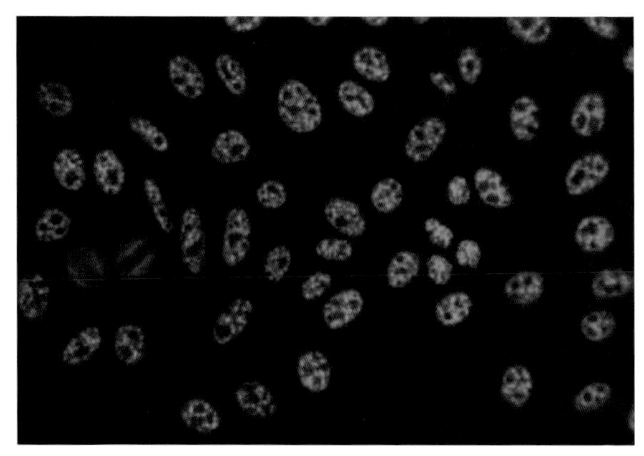

Sm/RNP antibodies. The interphase cells demonstrate a speckled nuclear stain that spares the nucleolar area. The metaphase chromosomes are unstained. In patients with antibodies to DNA or chromatin, the metaphase chromosomes may have homogeneous staining suggesting a mixed antibody pattern. Antibodies to Sm are a marker for SLE, while antibodies to RNP may be seen in various connective diseases.

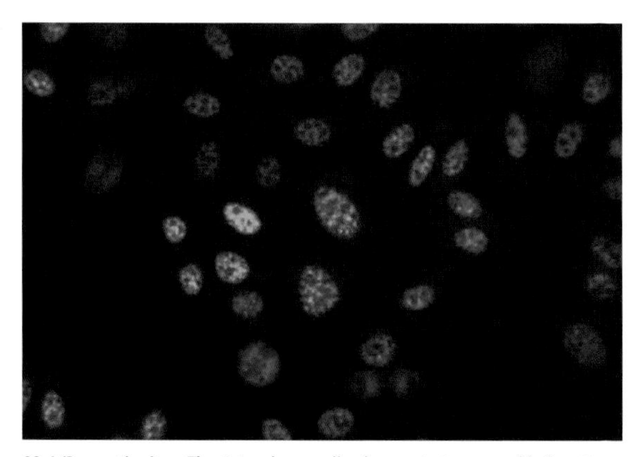

SS-A/Ro antibodies. The interphase cells demonstrate a speckled pattern. The metaphase chromosomes are unstained. These antibodies are present frequently in SLE, Sjögren's syndrome (SS) and neonatal lupus and with less frequency in other connective tissue diseases.

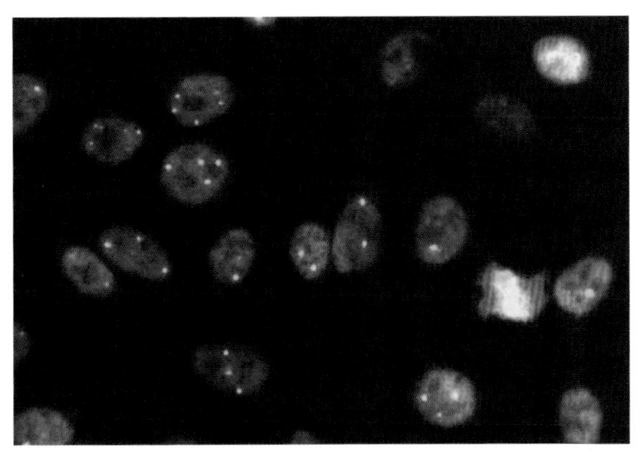

P-80 coilin antibodies. The P-80 coilin antibodies stain from 1 to 5 or 6 discrete speckles in the nucleus. The number of speckles is variable from cell to cell depending on the cell cycle. A monoclonal antibody to P-80 stains the discrete speckles only without any staining on the dividing cells. Frequently the P-80 antibodies are associated with a diffuse nuclear stain. These antibodies give a diffuse stain on the chromosomes. These antibodies have been reported in low frequency in patients with SS, SLE, and PBC.

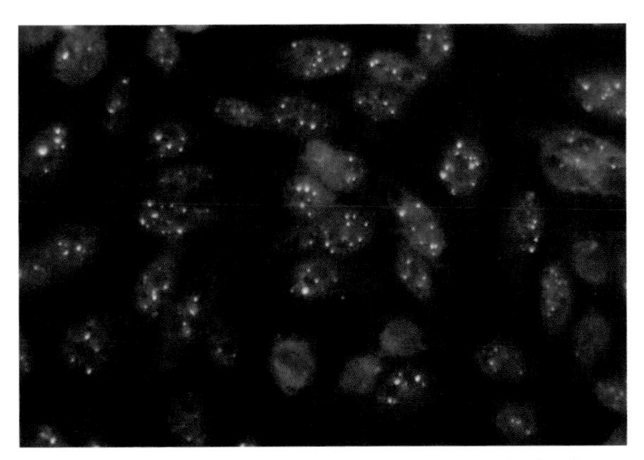

Nuclear dot antibodies. These antibodies give a nuclear stain with a few discrete speckles that vary in number from cell to cell. The metaphase chromosomes are not stained. The pattern has been seen with several different antibodies. There is no specific clinical significance.

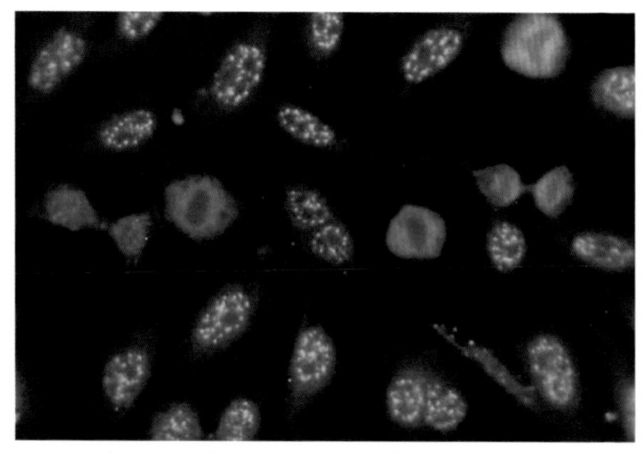

Coarse speckled antibodies. The interphase cells demonstrate many discrete speckles. The nucleolar area is spared. The metaphase chromosomes are unstained. This pattern may be confused with antibodies to centromere if there are no dividing cells to observe the metaphase area. There is no definite clinical association.

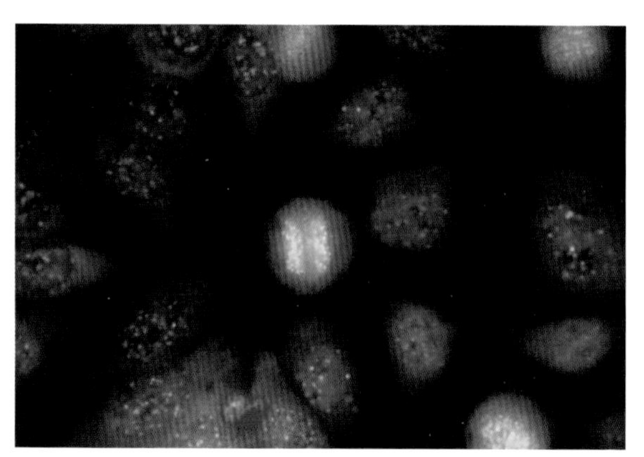

Centromere antibodies. The interphase cells stain in a discrete speckled pattern. The metaphase chromosomes demonstrate discrete speckles associated with the centromere areas of the chromosomes that changes with cell division. The cell in the center of the photo is in anaphase. These antibodies are generally associated with a limited form of PSS.

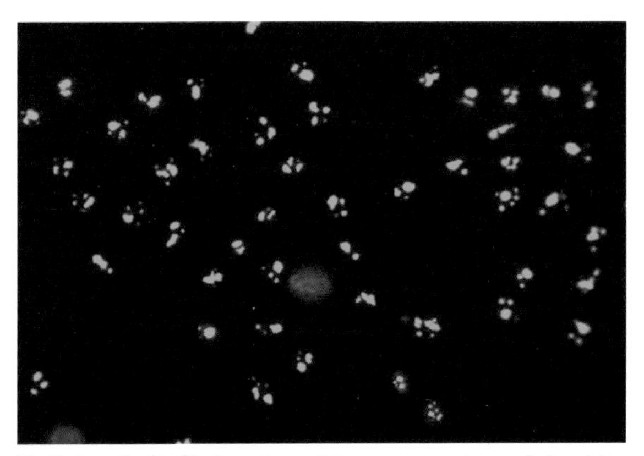

Fibrillarin antibodies. The interphase cell demonstrates a characteristic staining of the nucleolus. This has been described as a "clumpy" pattern. The dividing cells demonstrate a surface staining on the chromosomes. These antibodies are associated with scleroderma.

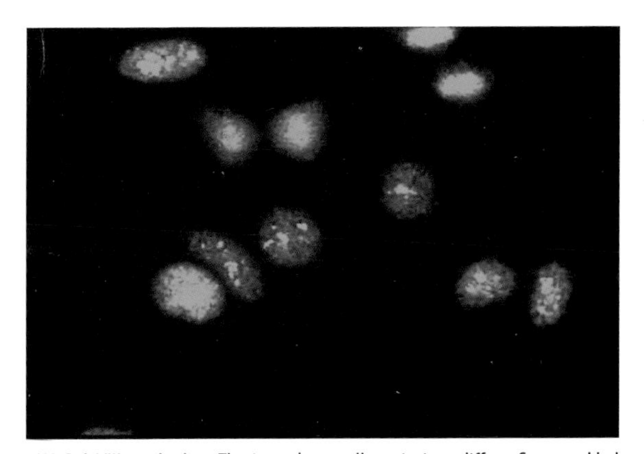

RNA Pol-I/III antibodies. The interphase cells stain in a diffuse fine speckled stain with the nucleolar staining appearing in a "speckled" distribution. This represents antibodies to RNA pol-I/III. The dividing cells, depending on the HEp-2 preparation, demonstrate a few discrete speckles associated with nucleolar organizing regions (NOR) of specific chromosomes. These antibodies are associated with scleroderma.

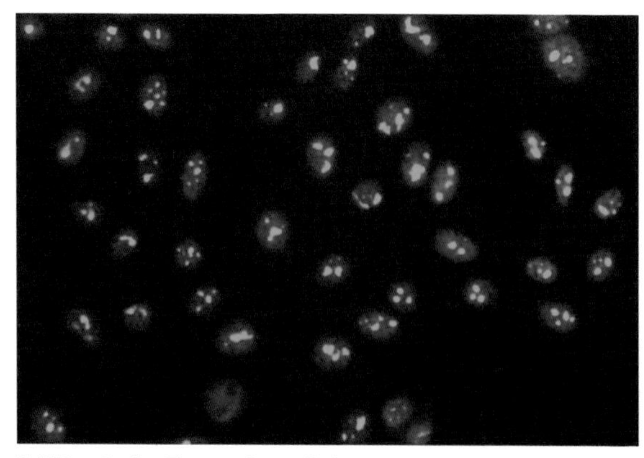

PM/Scl antibodies. The interphase cells demonstrate a homogeneous nucleolar pattern, with a less intense diffuse nuclear stain. The chromosomes of the dividing cells are unstained. This antibody was originally described in patients with a polymyositis/scleroderma overlap but may also be seen in patients with either polymyositis or scleroderma.

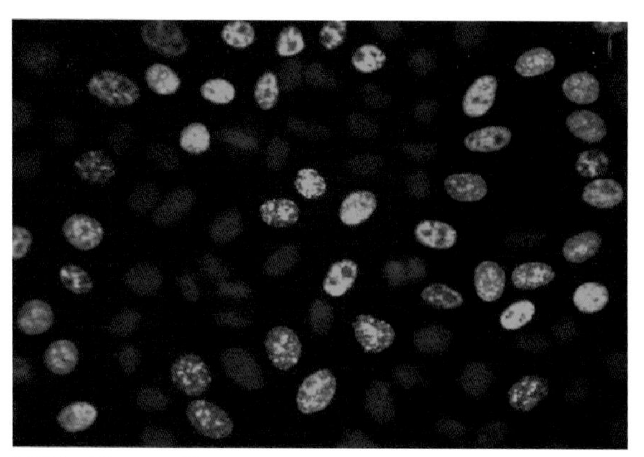

PCNA (proliferating cell nuclear antigen) antibodies. Some of the cells have very little stain and others demonstrate highly variable staining both in pattern and intensity. The brightly staining cells are in the S phase of the cell cycle. PCNA is an accessory protein for DNA polymerase delta that is up regulated in S phase. The metaphase chromosomes are unstained when PCNA is the only antibody present. There is a low incidence in SLE.

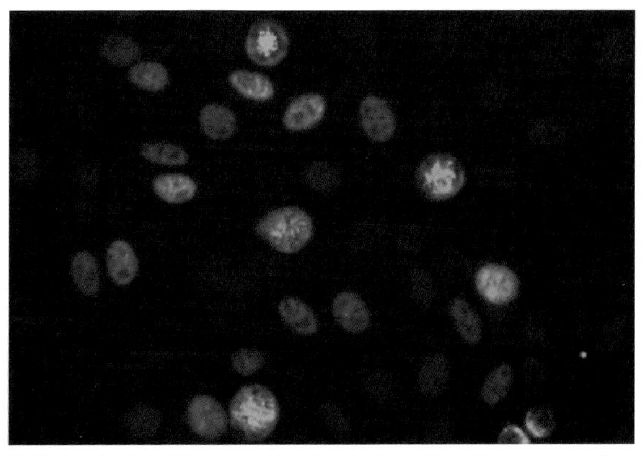

CENP-F antibodies. This pattern is cell cycle dependent. The majority of the cells have little stain. The cells staining are in G2/M phase of the cell cycle. The antigen distribution varies from diffuse in G2, centromere region at metaphase and intercellular bridging at cytokinesis. While it may be seen occasionally in patients with rheumatic disease, it is more frequently identified in cancer patients.

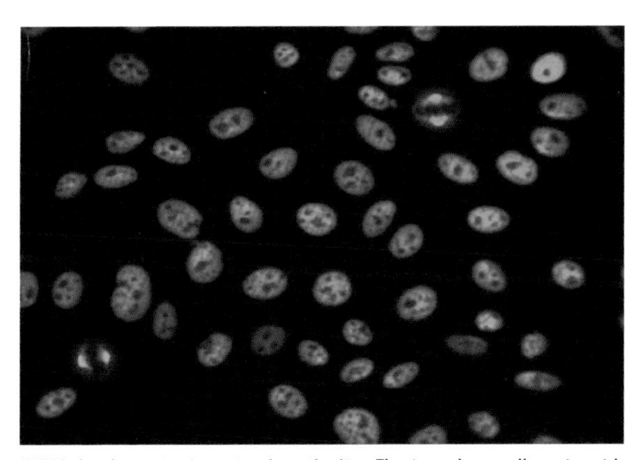

NUMA (nuclear-mitotic antigen) antibodies. The interphase cells stain with a diffuse fine speckled pattern. The metaphase chromosomes are negative but a staining of the spindle occurs. This is the most common but not the only antibody to stain the spindles. It is reported in Sjögren's syndrome and undifferentiated connective diseases.

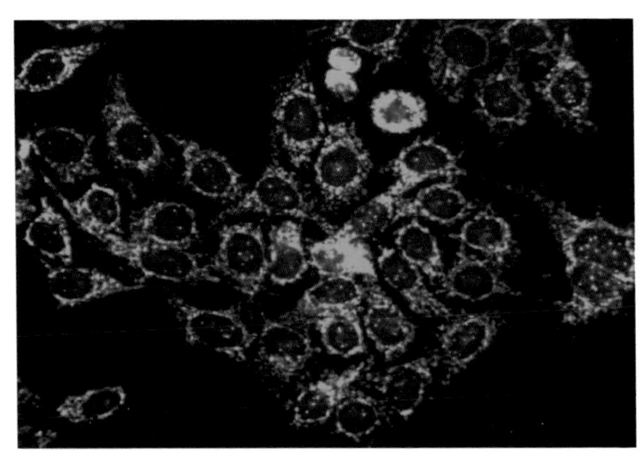

AMA (anti-mitochondrial antibodies) + Sp 100 antibodies. The AMA antibodies stain the cytoplasm with a coarse granular stain. The metaphase is unstained. While AMA appear alone in many cases, they are frequently associated with ANAs. The Sp 100 antibody in this picture is one of them. It is one of the antibodies reacting with nuclear dots. This pattern is seen in patients with PBC. Other ANAs associated with AMA in PBC include centromere and ones reacting with the nuclear membrane. When a strong homogeneous pattern occurs with AMA the diagnosis is more likely to be autoimmune hepatitis.

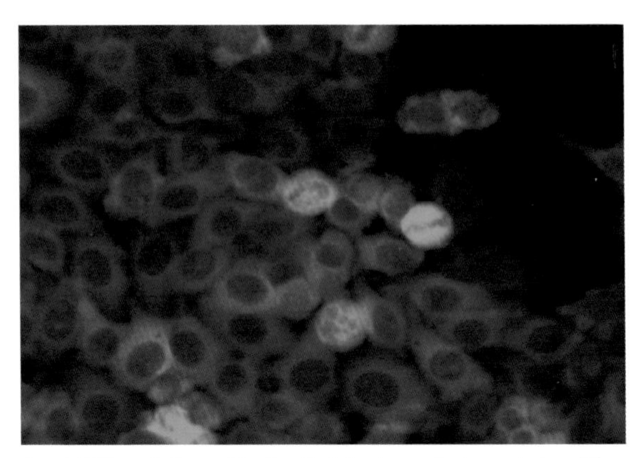

Alanyl-tRNA synthetase antibodies. The interphase cells demonstrate a diffuse cytoplasmic stain. The metaphase chromosomes are unstained. This is one of the tRNA synthetases that are detected in polymyositis. The most common antibody is Jo-1, which is ihnconsistently detected by IIF.

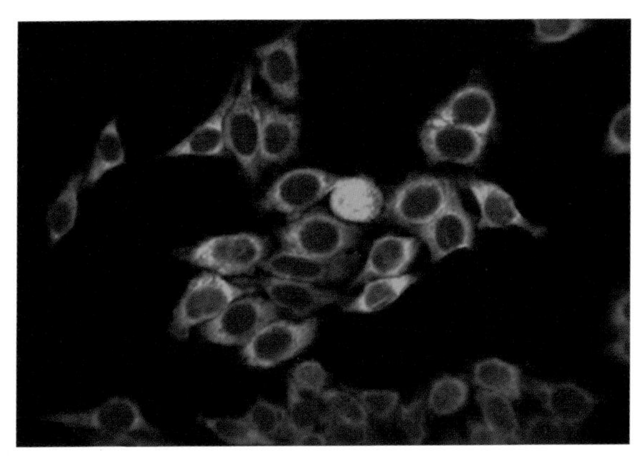

rRNP (ribosomal RNP, Ribo-P) antibodies. The interphase cells demonstrate diffuse fine speckled cytoplasmic staining. A nucleolar pattern is frequently associated. Since this antibody is primarily seen in SLE patients, it may occur with multiple other antibodies producing a mixed pattern. When occurring alone, the metaphase chromosomes are not stained. In the case of the mixed pattern, the metaphase reactivity will depend on the other antibodies present.

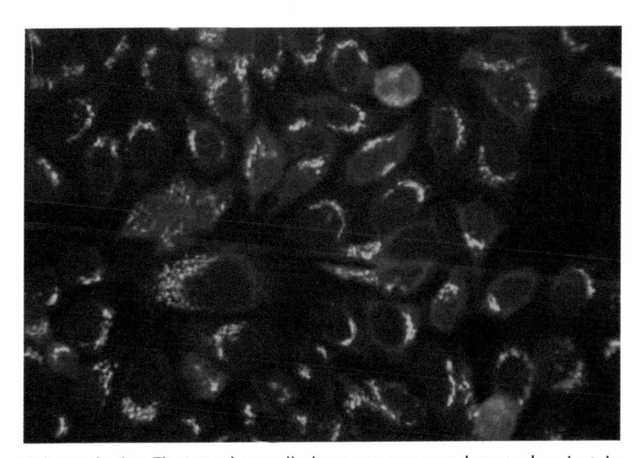

Golgi antibodies. The interphase cells demonstrate a granular cytoplasmic stain localized primarily on one side of the nucleus. The metaphase chromosomes are unstained. There is no specific clinical association. It has been reported in SLE, Sjögren's syndrome, undifferentiated connective disease. Other reports include idiopathic cerebellar ataxia, Neoplastic cerebellar degeneration and viral infections including HIV and EBV.

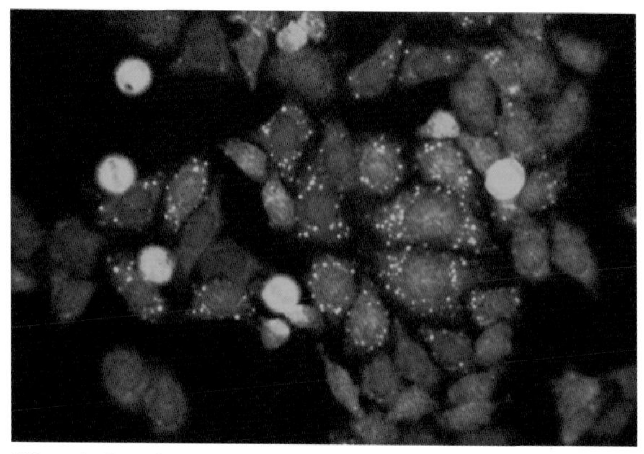

GW antibodies. The interphase cells demonstrate large speckles scattered throughout the cytoplasm. The metaphase chromosomes do not stain. This antibody may occur as a part of a mixed pattern. The antibody has been reported in patients with SS or with motor and sensory neuropathy.

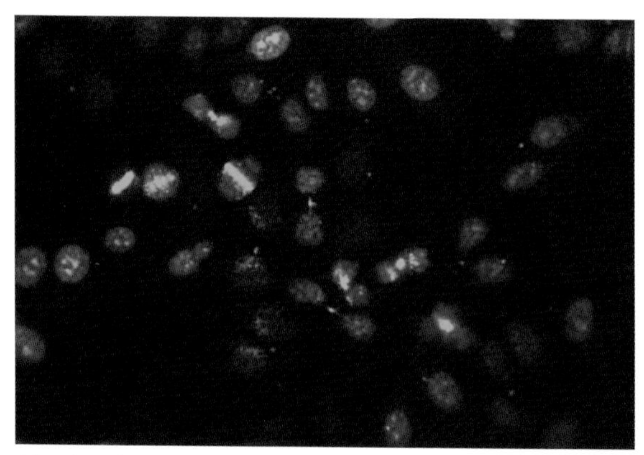

Cell cycle nucleolar and Stembody antibodies. This represents a mixed pattern. The cells demonstrate both a cell cycle nucleolar pattern and the stembody pattern. The stembody antibodies demonstrate a cell cycle pattern. They stain the metaphase chromosome area in a coarse banding pattern. In early cytokinesis they demonstrate a strong banding pattern at the constriction site between the two reforming nuclei and in late cytokinesis are seen as an intercellular bridge between the cells. The stembody pattern may occur alone or in other mixed patterns. The clinical significance is unknown.

Chapter 2

Identification of Antibodies to Cellular Components In Systemic Rheumatic Disease

Identification of Antibodies to Cellular Components In Systemic Rheumatic Disease

The antibody profile considered standard for the diagnosis of systemic lupus erythematosus (SLE), mixed connective tissue disease (MCTD), Sjögren's syndrome (SS), systemic sclerosis (SSc), and polymyositis/dermatomyositis (PM/DM) in most laboratories today includes the detection of antibodies to native DNA (nDNA), Sm, RNP, SS-A/Ro, SS-B/La, Scl-70, centromere and Jo-1. An antibody not always included in the standard ANA profile but one that is important to the diagnosis of rheumatoid arthritis (RA) is rheumatoid factor (RF). What follows is a brief overview of how it was discovered that these particular antibodies were useful in the diagnosis of systemic rheumatic disease (SRD).

The first assay employed for the detection of autoimmune diseases came from the observation of Waaler [12] in 1940 that sheep red blood cells sensitized with rabbit anti-sheep red blood cells would agglutinate in sera of patients with RA. He named the factor responsible for the agglutination Rheumatoid factor. RF, while detected in relatively high titer in patients with RA, was not specific for RA but could be detected in patients with other conditions also. RF was later determined to be an antibody reacting with the constant region of IgG.

The next autoantibody discovered was based on the observation of an atypical cell seen primarily in patients with SLE that was reported in 1948 by Hargraves et. al, [13]. He named it the LE cell. An *in vitro* assay was developed to detect the "LE cell factor". It was later demonstrated that the "factor" responsible for the formation of the LE cell was an antibody to nucleoprotein [14]. What Hargraves described was a cell with an unusual morphology detected in the blood and bone marrow of certain patients with SLE. The cell was a polymorphonuclear leukocyte

(PMN) that contained engulfed amorphous nuclear material. The amorphous material was from a damaged lymphocyte nucleus. While it was a welcomed addition for the diagnosis of SLE, it was a difficult test to perform. It was also detected in conditions other than SLE. The test itself required that a clot from a patient be mashed through a mesh screen to damage lymphocytes. The mixture was incubated to allow a reaction to occur. Blood smears were then made, stained and examined under the microscope to determine the presence of the LE cells. The complexity of the assay conditions led to considerable variation in the results. The test was labor intensive and due to the variability had to be performed on three successive days. This variability became understandable when it was later demonstrated that the test required four factors including antibodies to nucleoprotein, complement, nuclei opsonized by antibody and viable PMNs capable of ingesting the opsonized nuclei [15].

The next major breakthrough came when a test for indirect immunofluorescence was developed. Applying this technique to the sera from patients with SLE allowed the whole field of autoimmunity to develop. The development of immunofluorescence began in 1942 when Coons et al [16] found they could couple a fluorescent molecule directly to an antibody. They applied this first to specific antibodies for the detection of specific antigens [17]. Then in 1954 they reported the use of an indirect method of immunofluorescence for the detection of antibodies to viruses [18]. By 1957 three different groups of immunologists applied this technique to patients with SLE demonstrating that there were antibodies in patient's sera that would bind to nuclear antigens [19-21]. Within the next ten years it would be demonstrated that patients with different diseases gave characteristic patterns of immunofluorescence on cells and tissue sections [22].

During this time various investigators found a specific interaction of DNA with the sera from patients with SLE. A landmark observation was made in 1959 by Deicher et. al. [23] that established some important features of this reaction. They demonstrated that purified DNA from a number of species was equally reactive with SLE patient's sera and that the reactivity was due to an immunoglobulin. The clinical relevance of monitoring antibodies to DNA in the serum of SLE patients was reported by Tan et. al. [24]. In this paper the authors reported that antibodies to DNA were detected in the serum of a patient on four successive occasions prior to a flare of the disease. At the time of the flare antibodies to DNA were not detected. Instead, the patient had free DNA circulating in the serum. A follow-up study from the same laboratory [25] showed that immune complexes of DNA and anti-DNA could be eluted from the kidneys of some patients with lupus nephritis. The assay for detection of antibodies to DNA that is still considered the gold standard today was developed in 1968 [26].

Through the application of immunodiffusion techniques, antibodies to other nuclear antigens were soon observed [27,28]. The first major antibody to an extractable nuclear antigen was characterized in 1966 by Tan and Kunkel [29]. This was the Sm antibody. It was named after the patient used as a prototype, Smith. Naming antibodies using the first two letters of the last name of the prototype patient was a standard way for naming antibodies. The current method now relies on the identification of the specific protein or a designation of its size as determined by electrophoresis (Scl-70, P80 coilin, DFS-70). Soon after this, antibodies to RNP were described. Both anti-Sm and anti-RNP gave a speckled nuclear ANA but RNP antigen was differentiated from Sm antigen by its sensitivity to RNase [30-32]. Sharp et al [33] reported a rheumatic disease associated with high titer antibodies to RNP they named mixed connective tissue disease (MCTD) because these patients presented with symptoms of an overlap of SLE, PM/DM, and SSc. It was noted at that time that there was a linkage between the detection of antibodies to RNP and Sm.

It became apparent by the early 1970's through the application of indirect immunofluorescence (IIF), immunodiffusion, hemagglutination, and radioimmunoassay (RIA) techniques that a particular pattern of antibody reactivity was found in each of the systemic rheumatic diseases. In 1975 Notman, Kurata and Tan [31] first reported the use of "antibody profiles" for the diagnosis of SRD. This study tested for antibodies in patients with SLE, RA, SS, SSc, DM/PM, discoid LE, and MCTD to native DNA (nDNA), single strand DNA (ssDNA), chromatin, Sm and RNP. The results indicated that while antibodies to nDNA were detected in low titer in a number of the different diseases, they were detected in higher titers in patients with SLE. The chromatin antibodies were present almost exclusively in SLE patients. Antibodies to RNP were detected in many of the diseases while antibodies to Sm were limited to only SLE with one exception. That exception was a patient diagnosed with MCTD. Upon review, the patient was found to have SLE. This confirmed that even though antibodies to Sm are not a sensitive marker for SLE, they are specific.

Identification of antibodies to SS-A/Ro and SS-B/La associated with SS and SLE soon followed. Originally antibodies to both Ro [34] and La [35] were described as reacting with cytoplasmic antigens in patients with SLE. SS-A and SS-B were described as antibodies to nuclear antigens detected in patients with SS [36]. It was later determined that SS-A was the same as Ro and SS-B was the same as La [37].

The first extractable antibody defined in SSc was named Scl-1. The antigen was later purified and described as a basic 70 kD nonhistone protein [38]. At that time the name was changed to Scl-70. Patients with SSc were demonstrated by IIF to react with various nucleolar antigens giving three distinctive ANA patterns, homogeneous, speckled and clumpy [39]. The identification of the antigens responsible for these reactions occurred later. Also detected by IIF in patients with SSc, utilizing the new commercially available HEp-2 cells, were antibodies reacting with the centromeric region of chromosomes [40]. The clinical usefulness of differentiating SSc patients with these different antibodies became readily apparent.

With the identification of the antibody Jo-1 to the cytoplasmic antigen histidyl t-RNA synthetase in patients with polymyositis (PM) in 1982 [41], it was apparent that each of the major SRDs had an antibody profile characteristic for that disease. Patients with SLE were characterized by the presence of antibodies to nDNA, chromatin, Sm and RNP. Antibodies to SS-A/Ro, and occasionally antibodies to SS-B/La might also be present. Patients with SS demonstrated the combination of SS-A/Ro and SS-B/La. Patients with SSc could be differentiated by the presence of antibodies to Scl-70 or centromere. Many of the PM/DM patients demonstrated antibodies to the cytoplasmic antigen Jo-1. An overlap disease, MCTD, with characteristics of SLE, PM, and SSc, was characterized by high titers of RNP in the absence of other antibodies. Since this time many other antibodies have been identified in SRDs but they are not tested for routinely. For a review of the use of ANA profiles in SRD see the papers by Tan [9,10] and by von Mühlen and Tan [11].

The only major SRD without a characteristic antibody present was RA. While RF is present in many patients with RA, it also occurs in patients with other conditions. Numerous other antibodies were associated with RA but none were very sensitive or applicable to widespread clinical use. This changed in 1998 with the demonstration that many patients with RA react with certain citrullinated proteins [42]. An ELISA for the detection of antibodies to cyclic citrullinated peptides (CCP) was reported in 2000 [43]. The addition of CCP to the current profile now results in specific antibodies associated with each of the major SRDs.

As discussed above, numerous assays have been utilized to detect ENAs.

All of the clinically relevant autoantibodies to ENAs were detected using Oüchterlony double immunodiffusion using crude cellular extracts [5,6]. Later counterimmunoelectrophoresis (CIE) utilizing the same type of antigen was reported [32]. This increased the sensitivity of some of the reactions and decreased the time involved in obtaining results. Antibodies detected by these methodologies included Sm, RNP, SS-A/Ro, SS-B/La, Scl-70 and Jo-1. Antibodies to DNA were detected using RIA (26) or IIF of *Crithidia luciliae* [44]. Antibodies to chromatin were detected originally by the LE cell test, then RIA or a histone reconstitution assay and later by enzyme linked immunosorbant assay (ELISA) [45]. Antibodies to the centromere were described using IIF on tissue culture cells [40]. A hemagglutination assay was used to quantitate antibodies to Sm and RNP [46,47]. As the antigens were purified and eventually cloned, many were adapted for use in ELISA. This methodology is the one most commonly used at this time. It allows for the screening of large numbers of sera by automation producing an objective and semi-quantitative measurement of the antibodies detected. All of the above mentioned antibodies are available on ELISA. While ELISA is an excellent procedure, it requires each antigen to be tested individually. Newer technologies including the use of luminex beads and flow cytometry will allow for an even less labor intensive ability to screen samples since multiple antibodies may be measured in a single well instead of the multiple individual assays currently available in ELISA. Extensive discussions of the methods used to detect each antibody are found in *Autoantibodies* edited by JB Peter and Y Shoenfeld [2]. (See the technique section of the workbook for an overview of the specific procedures.)

The following presents an overview of the clinical aspects of the patients and the autoantibodies associated with the diagnosis of each of the major SRDs. The discussion of the assays for detection of the autoantibodies will focus on the types of antigens used, the secondary detecting reagents and other important variables and how they affect the final results. How to effectively apply FANA observations to the interpretation of specific antibody testing is also included. This should lead to a better overall understanding of autoantibody testing in general and the causes of variability between specific assays performed both within and between laboratories.

The two major causes of variation in assays come from the antigen itself and variations in the secondary antibody reagents used to detect the patient reactivity.

In order to understand how variations in an antigen can cause changes in an assay, it is important to understand the structure of an antigen and the ways that antibodies react with antigens. A majority of ENAs are protein in nature. A protein is composed of a characteristic string of amino acids. The order of the amino acids is called the primary structure. Because of the properties of the amino acids, the string of amino acids making up the protein in a short segment of the protein tend to fold back on themselves producing a simple secondary structure. The three dimensional structure of the entire protein is called the tertiary structure. When two or more separate macromolecules come together and interact tightly with each other the structure of the resulting complex is call the quaternary structure. (See Figure 1.)

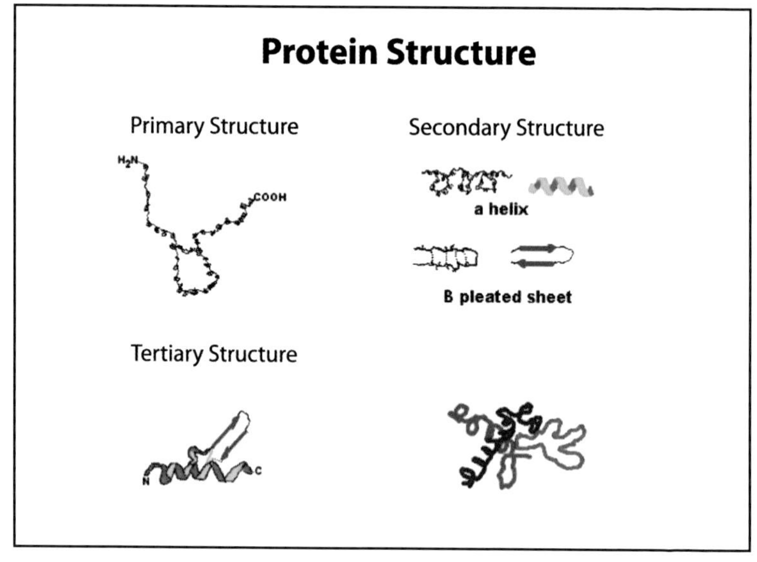

Figure 1. Protein Structure

The site that an antibody binds to is called an epitope. Antibodies are very selective in what they bind, and small changes in an epitope can cause large changes in antibody binding. The antibodies generated in an autoimmune response may be directed to any of the protein structures mentioned above. Specifically, they may recognize a linear epitope comprised of a few continuous amino acids (primary structure); a structure caused by local folding (secondary structure); a structure caused by global folding (tertiary structure); or a structure comprised of multiple macromolecules (quaternary structure). Antigenic sites created by protein folding or protein-protein interactions are termed conformational. Thus if proteins become unfolded thru purification or some other reason, some antigenic sites may be destroyed, changing the reactivity of the assay. An example is seen with antibodies to threonyl t-RNA synthetase, an antigen related to Jo-1, found in myositis. These antibodies can be detected by immunoprecipitation with radiolabeled antigen in its native form but not by immunoblot where the antigen becomes denatured on separation by SDS-PAGE. Also, if the reactivity is to a conformational site on a complex formed by the interaction of several different proteins or nucleic acids, the separation of this complex into its components will affect the results. Examples here include antibodies to chromatin and Sm/RNP. So, in the evaluation and comparison of assays, the nature of the antigen must always be taken into consideration.

Antibodies to all classes of immunoglobulins may be detected in SRD but the predominant and diagnostically significant antibody class is IgG. This is consistent with the recommendations of several authors to use IgG-specific fluorescent conjugates to detect ANA by IIF [48-50]. The reason for this is the result of the apparent antigen driven nature of the immune response. In a primary immune response the first antibody produced is of the IgM class. Over time, there is an isotypic class switch to IgG. The secondary immune response or continuation of the antigenic stimulation results in the production of primarily IgG. For this reason, most of the current assays measure IgG antibodies only. Antibodies to IgA and IgM are measured in specific instances only.

In comparing assay results it is important to recognize that assays vary in sensitivity and their ability to measure different types of antibodies. IgM antibodies are detected more efficiently in agglutination assays than are IgG. To enhance the reactivity of IgG antibodies in agglutination assays, the diluent must sometimes be modified. Because of the mobility of IgG immunoglobulins in an electrophoretic field, these are the antibodies detected in CIE. The ability to fix complement varies between the immunoglobulin classes as does their ability to bind to Staphylococcus protein-A. Discussion of some of these variables will follow in the sections on the diagnosis of the specific diseases.

Systemic lupus erythematosus

SLE is a multi-system disease affecting numerous organs and tissues. It is characterized by the presence of multiple autoantibodies directed primarily to nuclear and cytoplasmic antigens. Other antibodies detected include those to phospholipids, β_2-glycoprotein 1, and to cell surface antigens on platelets, RBCs, lymphocytes and neuronal cells. It affects predominately females between the ages of 15 and 40. The F/M ratio is 4:1 in early childhood increasing to 8:1 with age. It is more common among African Americans, Asians, and Hispanics. The prevalence in Caucasians is 1/2500 while in African Americans, Asians, and Hispanics is approximately 1/250 [51]. The exact numbers vary depending on the study. While some cases appear to have familial inheritance, most are considered to be spontaneous.

There are 11 criteria for the diagnosis of SLE. The presence of at least four criteria either serially or simultaneously is necessary to make the diagnosis. The criteria include malar rash, discoid rash, photosensitivity, oral ulcers, non-erosive arthritis, serositis, including pleuritis and pericarditis, renal disorder with persistent proteinuria or cellular casts, neurological disorder with seizures or psychosis, hematological disorders affecting RBC and/or WBC or platelets, immunological disorder with antibodies to nDNA, Sm or phospholipids, and a positive ANA [51]. The focus will be on the immunological abnormalities.

Most fluorescent ANA screening tests are ordered to "rule out lupus". With the use of appropriately fixed Hep-2 cells, the fluorescent ANA screen is very sensitive and it is extremely rare to find what could be called a case of ANA-negative lupus. The original reports of ANA-negative lupus came about because the substrate, mainly rat or mouse kidney, did not detect antibodies to SS-A/Ro [52]. In SLE, the major antibodies include nDNA, chromatin, Sm, RNP, SS-A/Ro, SS-B/La and ribosomal RNP (rRNP). All except rRNP react with nuclear antigens.

DNA

Antibodies to nDNA are considered one of the criteria for SLE. Antibodies to DNA may be to either the double stranded native DNA (nDNA) or to denatured single strand DNA (ssDNA). It is important to be able to discriminate between them. Antibodies to ssDNA are detected in numerous diseases but are not considered clinically relevant. Antibodies to nDNA occur in up to 70% of SLE patients with active disease. The detection of these antibodies in high titer is of clinical importance. Historically, antibodies to nDNA have been considered to be the major antibodies implicated in the pathogenesis of SLE. New research suggests that chromatin and not just DNA may play an important role [53]. Immune complexes containing both DNA and chromatin have been demonstrated in deposits in the kidney [53-56]. The deposition of these complexes may initiate an inflammatory response causing severe glomerular damage leading to renal failure. Historically the presence and /or increase in the titer of antibodies to nDNA with a concomitant decrease in complement are associated with a flare of disease, particularly renal disease. Physicians monitor the nDNA titers in SLE to try to predict flares so as to modify patient treatment.

Numerous assays have been developed to detect antibodies to nDNA. While there are a number of variables in DNA assays, the primary source of the antigen may have only a slight bearing on the results of the assay as most nDNA antibodies recognize the overall structure of DNA, not a specific sequence. Common sources of antigen include bacterial, plasmid or purified calf thymus DNA. What is important is how the antigen is presented and that the DNA does not contain single strand regions or protein contamination. The type of assay as well as the conditions of the reactions has a major effect on the results of the tests. Examples of 3 major assay types include RIA as in the Farr assay [26], IIF as in the *Crithidia luciliae* assay [44], and ELISA [45]. Comparisons of the different methodologies demonstrate that each detects antibodies the others do not [57]. Overall, most assays will agree about 85% of the time. No one assay will detect all nDNA antibodies.

The Farr RIA, still considered the gold standard, depends on the reaction of antibodies with radioactive DNA in solution and the precipitation of the antibody-DNA complex from the solution. The use of ammonium sulfate to precipitate the complexes leads to the detection of a subset of nDNA antibodies that react with nDNA in solution and are not dissociated in high salt. The presence of the high salt wash may account for differences when comparing these results to other assays. Occasionally proteins other than immunoglobulins will bind to the labeled DNA and be precipitated, leading to false positive results. An advantage of this assay is that antibodies detected are associated with renal disease in SLE patients. The major disadvantage, besides being labor intensive, is it requires the use of radioisotopes.

The anti-DNA test most widely used today is an IIF procedure using *Crithidia luciliae* as a substrate. It detects only antibodies bound to nDNA. The antibodies bound reflect those antibodies that will bind in a physiological salt solution in a solid phase format and not just those that remain bound in high salt solutions. *Crithidia luciliae* is a hemoflagellate with a large kinetoplast that contains circular nDNA. This circular nDNA has no ssDNA regions and no histones. Variables in the test include the growth conditions of the organism and the method of harvesting and fixing them on the slide. As with all IIF tests, the conjugate is important. The use of IgG specific conjugate will detect those antibodies associated with more severe SLE. Advantages of the *C. luciliae* assay include the simplicity of the assay, the ability to determine the class of immunoglobulin and the ability to determine if the antibodies fix complement. The disadvantages are that the test is labor-intensive and requires a trained technician to interpret the results.

The detection of anti-nDNA has also been adapted to ELISA. While the procedure for detection of nDNA is standard, the method to bind DNA to the plate and the reagents used in the assay vary greatly between commercial kits. Unlike protein antigens, nDNA will not bind directly to microtiter plates. It requires a linker molecule such as poly-L-lysine, protamine, methylated BSA or other types of molecules. The type of linker selected is extremely important since widely variable results will occur with the same antibody run at the same time using different linkers [58]. Other variables in the assay include the amount of antigen coated on the wells and the conjugate used to detect the bound antibodies. The international study [59] on evaluating ELISA methods for the detection of antinuclear antibodies found the tests for nDNA were among the most variable. Like the IIF, ELISA DNA assays do not employ a high salt buffer as in the Farr assay. They also present the DNA in a solid phase format and not in solution. For these reasons, they will detect antibodies not detected in the Farr and in some instances those not detected by IIF. On the other hand, the Farr assay detects some anti-nDNA antibodies that are not detected by ELISA.

For proper interpretation by the clinician, it is important to include the method used to detect the nDNA antibodies in the report. In comparing patients with discrepant results between the Farr assay and the *C. luciliae* assay, McGuigan et. al. found that those that were positive by only one of the assays were less likely to have SLE than if they were reactive in both assays [60]. Because of this, it may be of value to have more than one assay for DNA available in the laboratory. Remember, in talking to a clinician or discussing the differences that may occur between laboratories, discuss the type of assay employed, the nature of the antigen and particularly the detecting reagent used.

Chromatin

The LE cell discussed in the introduction was shown to be due to antibodies to nucleoprotein. The modern term for nucleoprotein is chromatin. Chromatin is comprised of 40% DNA, 40% histone and 20% non-histone protein. The histones are organized along the DNA as repeating units called nucleosomes. Each nucleosome is composed of a histone core containing 2 each of H2A, H2B, H3 and H4 wrapped around 2 times by 165 bp of DNA (See Figure 2.) with H1 on the outside. The nucleosomes are strung together by linker DNA. Antinucleosome or antichromatin antibodies react with DNA, the parts of histones that are exposed in chromatin, and conformational epitopes that are formed by the DNA/histone complex.

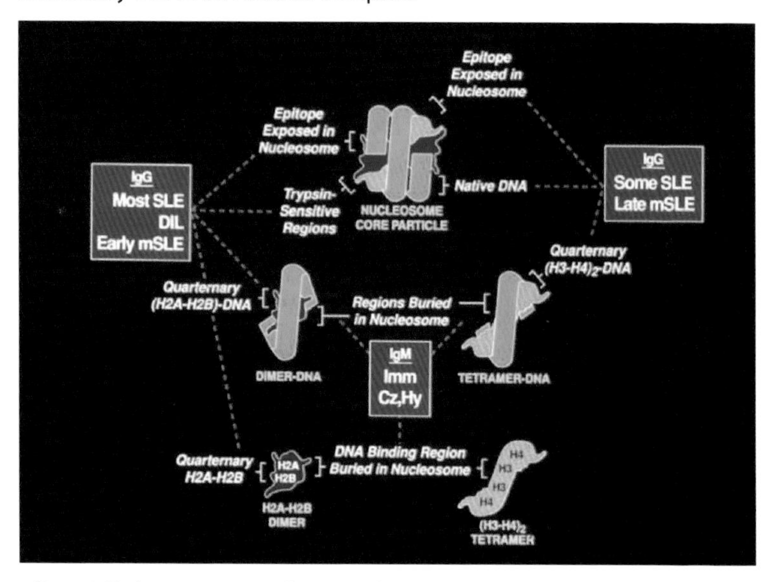

Figure 2. Nucleosome structure. (Courtesy of Dr. Rufus Burlingame, INOVA Diagnostics, Inc.)

In a mouse model of SLE it was demonstrated that antibodies to chromatin are the first antibodies to appear and that specific antibodies to DNA and histone develop later. The antibody profile suggests an antigen driven response with subsequent epitope spreading [61]. It is thought that this may also occur in human SLE. As mentioned in the section on nDNA, studies have demonstrated that antibodies to chromatin may also participate in the pathogenesis of renal disease in SLE as immune complexes isolated from the kidney contain chromatin as well as nDNA. Chromatin antibodies are detected in 50-80% of patients with SLE [62]. Antibodies to the individual histones are also detected in SLE. Antibodies to both histones and chromatin but not nDNA are detected in drug induced LE.

Methods to detect antichromatin and antihistone antibodies include LE cell, histone reconstitution, RIA and specific ELISA. Most laboratories use ELISA. The methodology is standard but, again, the antigens vary. The most useful antigen used to detect antichromatin/antinucleosome antibodies is H1-stripped chromatin [63]. This antigen is the core histone/DNA complex with H1 and non-histone proteins removed. It will detect antibodies to nDNA, the DNA-histone complex, and some of the histones. The antigen used in the assays for histones is a preparation of acid extracted histones resulting in denatured histones devoid of DNA. Since some of the antibodies are known to react with conformational epitopes, the antigen preparations and coating procedure as well as the buffers are important.

Sm and RNP

Antibodies to both Sm and RNP are detected in patients with SLE. Early studies noted their linkage [10]. In SLE it is common to detect patients with antibodies to RNP alone or antibodies to both RNP and Sm. It is rare to detect patients with antibodies to Sm alone.

Antibodies to Sm are detected in about 20-30% of SLE patients [11] while antibodies to RNP are detected in 30-40%. Sm antibodies are found almost exclusively in SLE and are considered a "marker" antibody. RNP antibodies, on the other hand, while frequently detected in moderate titer in SLE are, when present alone and in high titer, a marker for the overlap syndrome MCTD discussed below. RNP antibodies are also detected, usually at a lower titer, in other SRDs. Because of the clinical relevance of the titers observed in the SRDs, it became necessary not only to detect but also to quantitate the RNP antibodies in a patient's serum.

Sm and RNP antibodies, as well as the clinically relevant antibodies to ENAs, were first identified using Oüchterlony immunodiffusion [5,6,29]. (See the appendix for more coverage on the techniques.) The common sources of antigens are soluble extracts of either calf or rabbit thymus, or tissue culture cells. All of these sources contain the Sm and RNP antigens but may vary in the presence of other antigens, especially SS-A/Ro. While identification of the antibodies was achieved by immunodiffusion, it was not considered the method of choice for quantitation. Two important properties of the RNP antigen that allow for differentiation and subsequent quantitation of RNP and Sm are that the antigenic reactivity of RNP is destroyed by either treatment with RNase or by heat inactivation at 56°C for 30 minutes while that of Sm is not. (Because of this sensitivity to heat, care must be taken in the storage of antigens used for the detection of RNP.) Utilizing these differences, two methods for the quantitation of Sm and RNP were developed. These were hemagglutination in which both Sm and RNP antigens are adsorbed on to tanned sheep RBC and then the RNP reactivity is selectively removed by treatment with RNase [46,47], and counterimmunoelectrophoresis in which a portion of the cell extract is heat inactivated to destroy the RNP reactivity [64]. In both the assays the treated and untreated antigens are titered in parallel and the endpoint titers reported. Because agglutination and CIE differ in the types of antibodies preferentially measured, titers will vary between the methods.

In 1979 Lerner and Steitz [65], using sera from patients with SLE, demonstrated that antibodies to Sm and RNP both reacted with the same complex of proteins and small uridine rich nuclear RNAs. This RNA/protein complex was shown to function in messenger RNA splicing and was named the spliceosome. Using immunoprecipitation of radiolabeled proteins it was noted that both anti-Sm and anti-RNP brought down a characteristic complex of seven proteins designated A, B, C, D, E, F, and G. (See Figure 3)

Comparison of Sm and nRNP

Immunoprecipitation Sm	Immunoprecipitation nRNP		Immunoblotting Sm	Immunoblotting nRNP
		70 kD		—
—	—	33 kD-A		—
		29 kD-B'	—	
—	—	28 kD-B	—	
	—	22 kD-C		—
—	—	16 kD-D	—	
—	—	13 kD-E	—	
—	—	11 kD-F		
—	—	9 kD-G		

Ouchterlony

RNP ab

Sm ab

CTE

Figure 3. Comparison of Sm and RNP.

With the application of Western blotting, the difference between the protein antigens that were recognized by anti-Sm and anti-RNP could be determined. It was demonstrated that most Sm antibodies react with the B and the D proteins and occasionally with the E protein but also react to an additional protein designated B'. RNP antibodies, on the other hand, react with the A and C proteins and also with an additional protein designated 70 kD. It was noted early on that when a patient with antibodies to Sm is placed next to a patient with antibodies to RNP in Ouchterlony immunodiffusion a weak spur developed which would point in the direction of the RNP well. With these observations, an explanation of the spur formation was now possible. The extract used for testing contains both the Sm/RNP complex of proteins and also free SmD. The antibodies in both sera react with the complex. The patient with antibodies to Sm contains antibodies to SmD. The SmD protein is small, 16 kD. It will diffuse rapidly and can only be bound by the anti-Sm serum. The antibodies to the SmD in the Sm serum will react with the SmD in the complex but there will still be some available to bind to the free SmD diffusing past the complex, resulting in the spur. An up to date discussion of Sm and U1-RNP antibodies as well as other spliceosomal proteins can be found in a new book on SLE [66].

It can be seen from the above discussion that the composition of the antigen and the format of presentation can make a difference in the results. In the case of standard Oüchterlony immunodiffusion or CIE the antigen is usually a cell extract with a mixture of antigens with native conformation and some free or denatured antigen. The antigen for the immunoprecipitation assay is in solution and is usually native in structure. The antigens in the immunoblot are denatured during separation with some renaturation occurring during transfer to the solid phase.

Today ELISA has become the method of choice for identification of specific ENAs. The procedure is based on purified antigens being adsorbed on a solid surface. In the standard format the antigens are adsorbed onto 96 well microtiter plates but an alternative method is to apply the purified antigens on a nitrocellulose surface. This alternate method is designated immunostripe or line immunoassay (LIA). In either case the nature and integrity of the antigen will determine what epitopes are available for the antibodies to bind. An understanding of some of the differences in antigens will lead to a better understanding of assay results.

Antigens used for commercial assays vary considerably. They come from several different sources that includes natural sources such as animal tissues or organs, cloned antigens grown in either bacterial or insect cells or, in some instances, synthetic peptides. The method used and the choice of protease inhibitors included during purification will have a major effect on the integrity of the proteins obtained during the process. There are a number of different techniques available to purify the antigens including standard biochemical and gel chromatography techniques as well as various forms of affinity chromatography. Depending on both the source and the purification techniques, the antigens will contain more or less tertiary and quaternary structure.

The use of cloned antigens introduces several variables not encountered in the use of natural antigen sources. Recombinant proteins are individual proteins, not complexes of multiple proteins and/or nucleic acids. Therefore, when cloned antigens are used, antibodies to epitopes created by the interaction of the different proteins with each other or with nucleic acids will not be detected. During protein translation, in both prokaryotes and eukaryotes, a linear polypeptide is produced with some areas of folding that are stabilized by what is termed a chaperone protein. When translation is finished, the polypeptide is released from both the ribosome and its chaperone protein and is then able to complete folding into its correct three-dimensional conformation. In eukaryotic cells further modifications often occur to form the final protein. These modifications, termed post translational modifications, include cleavage of portions of the polypeptide to form the "active" protein, modification of specific amino acids through the addition of methyl, oxygen or phosphate groups, or the addition of carbohydrates and lipids. In prokaryotic cells, post translational modifications do not occur. Thus, recombinant proteins propagated in a prokaryote host such as *E. coli* contain no post translational modifications while those propagated in a eukaryote host such as Sf9 insect cells do. Autoantibodies from patients usually react with multiple epitopes on the same protein. As a specific protein is composed of different regions, some of which are unmodified and others that are post translationally modified, antibodies may be present in a serum that react with epitopes in unmodified regions of the antigen and others that react with epitopes with post translational modifications. The use of an antigen without PTM would detect only the subset of the antibodies reacting with the unmodified regions of the protein. It should be noted that during the purification of recombinant proteins a small amount of the host protein may contaminate the preparation. As many individuals have antibodies to *E. coli,* any residual bacterial antigens remaining after purification could bind to the solid phase and lead to false positive results. With Sf9 cells this is not a problem because antibodies to insects are very rare. Thus, it is clear that the source of the antigen is an important variable in ELISAs.

From the discussion above, knowing that Sm and RNP are part of an RNA/protein complex and that antibodies in patients react with numerous epitopes on the specific proteins, one can appreciate the fact that different assays sometimes give different results. In commercially available ELISA kits some manufacturers coat the microtiter plates with the Sm/RNP complex and free Sm using biochemically separated antigens while others coat the plates with cloned antigens of differing components. This obviously leads to different antigens being available for reactivity and variable results. This must always be taken into consideration in comparing discrepant results between kits.

The ANA Standardization Subcommittee of the Standardization Committee of the International Union of Immunological Societies did a critical evaluation of ENA ELISAs. This study compared ELISA kits from 9 companies. The samples tested were prepared and coded at CDC using the CDC reference sera [67,68]. The antibodies tested include nDNA, ssDNA, histone, Sm, U1-RNP, SS-A/Ro, SS-B/La, Scl-70, centromere, and Jo-1. Not all 9 companies had assays for all of the antigens. Most of the companies did quite well with their detection of antibodies to SS-A/Ro, SS-B/La, Scl-70, and Jo-1 but demonstrated considerable variability in detection of antibodies to Sm and nDNA [59].

SS-A/Ro and SS-B/La

Antibodies to SS-A/Ro and SS-B/La are detected in SLE but, when they occur together, are more characteristic of patients with primary SS (See below for a discussion of SS). SS-A/Ro antibodies have an incidence in SLE of about 35%. The percentage varies between studies and depends on the assays used. SS-A/Ro antibodies are also associated with subacute cutaneous lupus, lupus of homozygous complement deficiencies, and the neonatal lupus syndrome. The incidence of SS-B/La in SLE is 10-15%. When antibodies to SS-B/La are present, the patients generally have sicca features [11].

Neonatal lupus syndrome is a rare condition observed in infants that is characterized by dermatological, cardiac or hematological abnormalities. The cause appears to be passive transfer of maternal IgG antibodies to the fetus. Antibodies to SS-A/Ro are detected in 75-95% of cases. SS-B/La antibodies are present 75% of the time [11, 69]. Most infants present with only dermatological changes but a few also develop congenital heart block that may be fatal. The overall risk for women with SLE is < 5%. In women with SLE and antibodies to SS-A/Ro the risk is about 15% that the baby develops dermatological manifestations. The incidence of congenital heart block is about 1-2%. It has been demonstrated that both the 52 kD SS-A/Ro and SS-B/La are abundant in the fetal heart between 18-24 weeks [70]. As some forms of these antigens are expressed on the cell surfaces, it is postulated that the maternal antibodies cross the placenta and react with the antigens on the surface membranes of the cells [71].

The SS-A/Ro antigen is variably expressed in different tissues [72]. Antibodies to SS-A/Ro are not detected by IIF using rodent tissue as substrate [52]. Because of this, many patients with these antibodies were diagnosed as having ANA negative SLE. With the use of acetone fixed HEp-2, these patients are ANA positive. It is rare to find a true ANA negative SLE patient.

Like antibodies to Sm and RNP, antibodies to SS-A/Ro 60 and SS-B/La show a linkage relationship. While anti-SS-A/Ro is frequently detected alone, anti-SS-B/La is rarely detected without anti-SS-A/Ro. Historically, three major proteins serve as antigens in this system. They include the SS-A/Ro 60, the SS-A/Ro 52, and the 48 kD SS-B/La. The SS-A/Ro60 and the SS-B/La both bind directly to small cytoplasmic RNAs designated hY1--hY5. These small RNAs are different from the u-RNAs bound by Sm and RNP. In addition, the SS-B/La antigen binds to nascent RNA polymerase III transcripts. New evidence suggests that the SS-A/Ro 52 is not a part of the complex. The SS-A/Ro52 antigen does not appear to bind directly to SS-A/Ro60, SS-B/La, or the hY1--hY5 RNAs [73]. It has been identified as TRIM 21, a member of the tripartite motif family [74]. It functions as an E3 ubiquitin ligase. [75,76] Newer studies indicate that antibodies to SS-A/Ro 52 are not detected in immunodiffusion and give a negative ANA. These antibodies are not limited to SS but are detected in a number of different diseases. Anti-SS-A/Ro 60 is more specific for SLE and SS than anti-SS-A/Ro 52.

Antibodies to SS-A/Ro and SS-B/La are detected in the laboratory by immunodiffusion, CIE, immunoblot, ELISA and LIA. As with the previous antibodies, there are antibodies to linear and conformational epitopes. The type of antigen used in an assay makes a difference in its sensitivity. In immunodiffusion and CIE the antigen is in it's native conformation in a cellular extract. In immunoblot/LIA and ELISA the antigens used may be affinity purified, recombinant, or a mixture of both which may contain denatured antigens. Detection of antibodies to SS-B/La is seldom problematic. This is not the case with antibodies to SS-A/Ro. Correct interpretation of the results depends on the source and presentation of the SS-A/Ro antigens. Some commercial ELISA preparations contain predominately the 60 kD protein and others contain both the 60 and the 52 kD proteins.

Antibodies to SS-A/Ro 52 were first observed by Ben-Cheterit et. al in 1988 while examining the autoantibodies present in patients with SS by immunoblot. They demonstrated that the SS-A/Ro 60 and SS-A/Ro 52 were distinct proteins and that the antibodies against them did not cross react [77].

Further studies examining patients with SS and SLE with antibodies to SS-A/Ro by immunodiffusion found that not all immunodiffusion positive patients were reactive by immunoblot [78]. But of the sera that did react, most patients with SLE or SS contain antibodies to both the 52 kD and 60 kD SS-A/Ro proteins but some do not. While some of the discrepancy was due to differences in antibody specificity, some was probably due to the fact that a major epitope of SS-A/Ro 60 that is recognized by autoantibodies is conformational. These antibodies do not bind to the denatured antigen on the gel. In 1997, Rutjes et.al [79] demonstrated that the antibodies to SS-A/Ro 52 were more common than the SS-A/Ro 60 in patients with polymyositis with antibodies to the tRNA synthetases. In their case antibodies to SS-A/Ro 60 and SS-B/La were detected in 4% of the myositis cases while those to SS-A/Ro 52 were detected in 20%. If one looked at the myositis patients with antibodies to Jo-1, the incidence of antibodies to SS-A/Ro 52 was 58%. In a multi-center study evaluating ELISAs specific for SS-A/Ro 52 and SS-A/Ro 60 (personal observations), the results indicated that the antibodies to SS-A/Ro 52 were the most frequently detected antibodies in polymyositis patients (See Figure 4). Even though antibodies to SS-A/Ro 52 may be detected in several diseases, there is an advantage to differentiating the SS-A/Ro components in patients suspected of having polymyositis. A new ELISA for screening for polymyositis containing Jo-1 (histidyl), threonyl and alanyl-t-RNA synthetases and SS-A/Ro 52 is commercially available.

SS-A/Ro Antibodies

	SS-A/ Ro 60 + 52	SS-A/ Ro 60 only	SS-A/ Ro 52 only
SLE (161)	28%	9%	7%
SjS (72)	74%	2%	12%
SSc (122)	7%	2%	19%
PM/DM/OL (199)	12%	2%	45%
tRnaS+ (82)	12%	1%	76%
Others (84)	1%	1%	6%

Figure 4. Importance of SS-A/Ro 60 and SS-A/Ro 52 antibody differentiation

Ribosomal Proteins

Antibodies to the ribosomal-P proteins (ribo-P) are the most frequently detected antibodies to ribosomal proteins. They are present in 10-20% of patients with SLE. While it has been reported that these antibodies are associated with psychotic symptoms in SLE [11], other researchers associate them with renal symptoms [80,81]. Antibodies to ribo-P stain predominantly the cytoplasm of cells with some nucleolar stain. Antibodies to ribo-P react with three proteins of the large ribosomal subunit. They are 38 kD, 16 kD, and 15 kD [82]. Antibodies to other ribosomal proteins or to 28S RNA are occasionally detected in SLE sera [83]. While ELISA is the most common assay to detect anti-ribo-P today, they were originally detected by immunodiffusion, immunoprecipitation, and immunoblot. The immunodiffusion, immunoprecipitation, and immunoblot assays use cell extracts for the source of the antigen. Some ELISAs use affinity purified native ribo-P while others use a peptide derived from the C-terminal portion of the molecule. This is one of the rare cases where a peptide is a useful antigen to detect autoantibodies.

Proliferating Cell Nuclear Antigen (PCNA)

Antibodies to PCNA recognize a cell cycle related nuclear antigen in proliferating cells [84,85]. This antibody is detected in 3% of unselected SLE patients. PCNA is an accessory protein to DNA polymerase δ that functions in DNA synthesis. It therefore is present in areas of the nucleus where DNA is being synthesized. This occurs during the S phase of the cell cycle. The ANA pattern observed during S phase reflects the uneven synthesis of the DNA resulting in a variable pattern. Those cells not in S phase will demonstrate little to no staining. While the ANA pattern can be characteristic of PCNA, the confirmation must be done by other methods. This is usually performed in specialty or research labs.

Other antibodies

Other antibodies reported in SLE are reactive against HMG-17, hnRNP protein A1, Ku, cardiolipins, and $ß_2$glycoprotein 1 [11].

Mixed connective tissue disease

MCTD is an overlap syndrome with symptoms of SLE, SSc, and PM and less commonly a destructive arthritis similar to RA [33]. Common features include Raynaud's phenomena, puffy or edematous hands, arthritis involving the small joints of the hands, and serositis. Patients may develop sclerodactyly and esophageal dysfunction. They usually do not develop the glomerulonephritis or CNS symptoms associated with SLE. One of the diagnostic criteria is the presence of antibodies to RNP in high titer with no other ENA antibodies being detected. These patients frequently have high titers of antibodies to RF.

Sjögren's Syndrome

Patients with primary Sjögren's syndrome (SS) present with dry eyes, dry mouth, myalgias, joint aches, fatigue, and depression. SS patients appear to be at an increased risk to develop lymphomas or pseudolymphomas and some patients may develop a mixed cryo-globulinemia containing an IgM-k monoclonal RF similar to those observed in Waldenström's macroglobulinemia. The exact percentage varies depending on the reports. Other systemic symptoms occur less frequently. Patients with secondary SS have the same symptoms as those with primary SS but also present with another CTD such as RA, SLE, PM, or SSc. The disease occurs predominately in females with an estimated incidence of 1: 1,250. The major antibodies associated with primary SS are SS-A/Ro and SS-B/La. Other antibodies include p-80 coilin and MA-1. The incidence of the antibodies to SS-A/Ro and

SS-B/La are 70% and 50% respectively [72]. High titers of rheumatoid factors are frequently detected [86,87].

Systemic sclerosis

Patients with SSc often present with a diffuse puffiness of the hand, carpal tunnel syndrome, Raynaud's phenomenon, and GI symptoms. The essential clinical parameter is skin thickening proximal to the metacarpophalangeal joints. It is a rare disease seen more frequently in women than men (3.8 to 1). The presentation may be localized or systemic. Overlap syndromes also occur. The systemic forms of the disease may be diffuse or limited. The diffuse form of the disease affects not only the skin but also many internal organ systems. These include pulmonary, renal, cardiac, GI etc. The limited form is known as CREST. This stands for calcinosis, Raynaud's phenomenon, esophageal hypomotility, sclerodactyly, and telangiectasia [88]. The major diagnostic antibodies associated with SSc are anti-Scl-70 and anti-centromere. There are also antibodies to several nucleolar antigens including RNA Pol I, fibrillarin, PM/Scl, and Th/To. Antibodies to RNA Pol II and Pol III are also detected. Diffuse scleroderma is characterized primarily by the presence of antibodies to Scl-70 (DNA topoisomerase I) and several nucleolar proteins. Limited scleroderma is characterized primarily by the presence of antibodies to the centromere [89]. Patients presenting with features of both SSc and PM may have antibodies to PM/Scl [11].

Scl-70

Scl-70 was first identified as a common precipitin band detected in patients with SSc when tested by immunodiffusion using rabbit thymus extract. The antigen was identified as a 70 kD protein in 1979 [38]. In 1986 three different groups identified the protein as DNA topoisomerase I, a 100 kD protein. [90-92]. What accounts for the difference in the size of the actual protein and the protein defined as Scl-70? Topoisomerase I is a labile protein that is readily broken down by endogenous proteases to multiple small peptides, the most stable of which is 70 kD. It is apparent from this that the protein described in the original paper was the stable breakdown product. Antibodies to Scl-70 have been reported to occur in 20-75% of cases of diffuse scleroderma depending on the study. It is considered to be a marker for SSc [93]. Current methods for detecting antibodies to Scl-70 include immunodiffusion, immunostripe/LIA, and ELISA.

Centromere

Antibodies to centromere are detected primarily by the distinctive pattern observed in IIF. The antibodies give a finite number of punctate speckles in the nucleoplasm with staining in the centromeric region of the condensed metaphase chromosomes. It was proven to be the centromeric regions themselves by staining chromosome spreads [40]. Further studies have demonstrated that the sera contain antibodies to at least 3 different proteins associated with the centromere designated CENP-A (17 kD), CENP-B (80 kD) and CENP-C (140 kD) [94]. These proteins have been cloned and are used in ELISA and immunoblot assays.

Nucleolar

In 1982 studies of SSc patients using HEp-2 cells demonstrated three different patterns of nucleolar ANA. These were speckled, homogeneous, and clumpy [39]. The speckled nucleolar was later identified as being predominately antibodies to RNA Pol-I [95]. The homogeneous nucleolar pattern is observed with numerous antibodies including PM/Scl and Th/To [96-100]. The clumpy nucleolar pattern was identified as fibrillarin [101,102]. The important point is they were all seen in SSc or in the case of PM/Scl in an overlap with PM. A recent article in *Science* [103] noted the current number of proteins identified from the nucleolus is 271.

As noted from the various patterns observed by IIF, the nucleolus is not a homogeneous organelle. It is composed of three major regions. They are the fibrillar center, the dense fibrillar component surrounding the fibrillar centers, and the granular component (See Figure 5). Antibodies giving a speckled nucleolar pattern are localized primarily in the fibrillar centers, those, like fibrillarin, giving a clumpy nucleolar

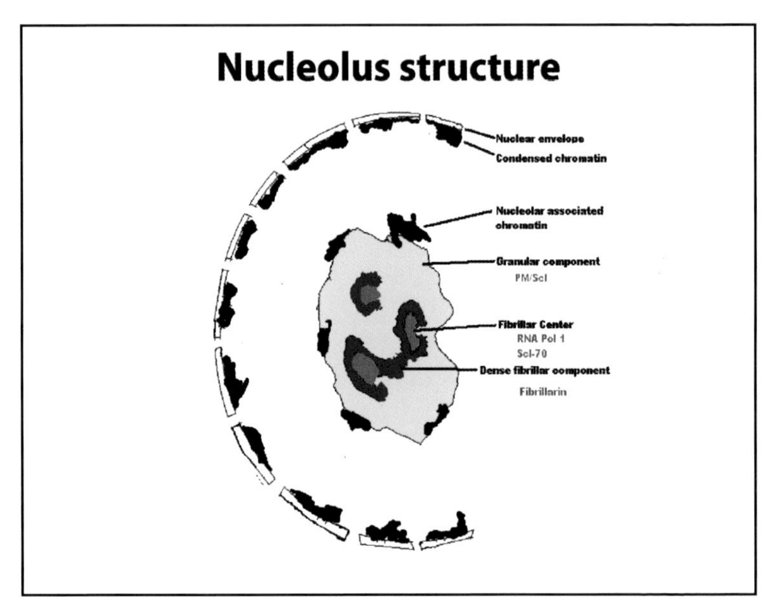

Figure 5. Nucleolus structure.

pattern are found in the dense fibrillar component, and those that are homogeneous react with antigens in the granular component.

Fibrillarin, the main protein in U3-RNP, is a 34 kD protein localized in the dense fibrillar component of the nucleolus. It is responsible for the "clumpy" nucleolar pattern. The name is derived from its location in the nucleolus. Although present in only about 8% of scleroderma sera [11], it is reported to be a marker for severe SSc especially in blacks and males [104-106].

Th/To is a 40 kD protein localized in the granular component of the nucleolus. Antibodies to Th/To stain the nucleolus with a homogenous appearance. Anti-Th/To may be detected in as many as 19% of patients with limited scleroderma, 11% with diffuse scleroderma, and 3% with primary Raynaud's disease [107,108].

Antibodies to PM/Scl, originally designated PM-1, were detected as a weak precipitin line by immunodiffusion in 61% of 28 PM/DM patients including 7/8 in a scleroderma/myositis overlap [109]. Later studies indicated more than one antibody was being detected. Sera were exchanged between several laboratories and the unique specificity of PM/Scl was defined [110]. While a few patients with either SSc or PM have antibodies to PM/Scl, about 50% of those with the antibody have PM/Scl overlap. However, in testing patients presenting with PM/Scl overlap, only 25% have the antibody. The outcome of patients with PM/Scl antibodies is favorable. The antigen is located in the granular component of the nucleolus resulting in a homogeneous pattern by ANA. The antigen is a complex of 11-16 proteins as determined by immunoprecipitation in SDS-PAGE [97]. Two of the proteins have been cloned. This includes the major band of 100 kD and a second, less prominent band of 70 kD. The two bands are immunologically distinct and the antibodies against one do not cross react with the other [111,112]. Antibodies to PM/Scl are currently detected by immunodiffusion or immunoblot/LIA in specialty or research labs.

Antibodies to RNA polymerases (RNAP) I, II, III are detected in scleroderma. RNAP-I is located in the fibrillar centers and is responsible for one of the " speckled" nucleolar patterns. RNAP II and III are located in the nucleus and antibodies to them give a speckled pattern on IIF. RNA polymerases catalyze the transcription of unique sets of genes. RNAP I transcribes ribosomal RNA genes, RNAP II transcribes all protein-coding genes, and RNAP III transcribes genes that produce small stable RNAs including 5S and t-RNAs. Each polymerase is a complex of 8-14 subunits ranging in size from 10-220 kD. Each polymerase has 2 distinct large subunits >100 kD and multiple smaller subunits, some of which are shared. Antibodies to RNAP I and RNAP III occur together as anti-RNAP I/III. Anti-RNAP I/III is highly specific for scleroderma. These antibodies are reported to be the most common SSc-related antibody in white North American patients [113]. They have a reported incidence of 23% [114] and are associated with diffuse cutaneous involvement, a high frequency of renal crises, and high mortality [115]. RNAP-II antibodies, while most often associated with SSc, are also detected in SLE and some overlap conditions. In a study of 32 SSc patients with antibodies to RNAP, 5 had antibodies to RNAP I/III and II, 23 had antibodies to RNAP I/III only, and 4 had antibodies to RNAP II only. The 4 patients with antibodies to RNAP II also contained antibodies to Scl-70. Patients with the latter combination of antibodies to both RNAP-II and Scl-70 were reported to be more often males with an older onset of disease, to have a greater frequency of renal and cardiac involvement and to have a reduced five-year cumulative survival. Detection of antibodies to RNAPs was primarily done as a research only test. The cloning of an immunodominant epitope on RNA Pol III [116,117] has allowed the development of a commercial assay for the detection of RNAP III.

Polymyositis/Dermatomyositis

Polymyositis and dermatomyositis are just two of ten classifications of idiopathic inflammatory myopathies. They are uncommon autoimmune diseases affecting the muscles that occur in 1 to 9 patients/million. They are more common in females than males (2:1) especially during the childbearing years (5:1). Patients with PM/DM generally present with symmetrical proximal muscle weakness affecting both upper and lower extremities. An elevated creatine kinase is detected in 95% of patients at some time during their disease. Criteria for the diagnosis of PM/DM include abnormal electromyographs (EMGs), and muscle biopsies demonstrating degeneration, regeneration, necrosis, phagocytosis, and an interstitial infiltrate. In the case of DM, also found are the characteristic heliotrope rash, a violaceous discoloration of the upper eyelids, or Gottron's sign, a scaling erythematosus eruption over the knuckles and elbows. The peak onset is between 40-60 years. Interstitial lung disease occurs in 10-30% of PM patients [118].

Autoantibodies detected in PM/DM can be divided into three different groups. The classification includes antibodies that are myositis specific, those that are myositis associated, and those that have been detected in myositis but their specificity has yet to be established. The myositis specific antibodies include antibodies to the t-RNA synthetases, signal recognition particle (SRP) and Mi-2. The myositis associated antibodies include PM/Scl, U1-RNP, U2-RNP, Ku, SS-A/Ro, and SS-B/La. The antibodies with specificity to be established include antibodies to MJ, nuclear pore complex, 56 kD protein, and U5RNP [110].

Patients diagnosed with anti-synthetase syndrome, a subset of PM, present with myositis, interstitial lung disease that may become severe, acute or fatal, arthritis, Raynaud's, fever, myalgias, and mechanic's hands. Their sera are positive for myositis specific anti-bodies. The syndrome occurs in 25-30% of myositis patients. The most commonly detected antibody is anti-Jo-1 or anti-hystidyl-t-RNA synthetase [119] that occurs in 18-20% of PM patients. Antibodies to other t-RNA synthetases include those to threonyl, alanyl, isoleucyl, glycyl and asparaginyl. The function of t-RNA synthetases is to catalyze the binding of an amino acid to its specific t-RNA during protein synthesis. As this occurs in the cytoplasm, staining on Hep-2 is cytoplasmic and not nuclear. Antibodies to Jo-1 are detected using ELISA, immunoblot/LIA, or immunodiffusion. A new ELISA screen for myositis detecting histidyl, threonyl, alanyl and SS-A/Ro 52 is currently available.

Of the myositis associated antibody group, most have been discussed previously. These antibodies, while detected in some patients with PM/DM, are not specific for PM/DM. Antibodies to U1-RNP are seen in MCTD, those to PM/Scl are seen frequently in the scleroderma/myositis overlap, and those to Ku, while being associated with PM, are also detected in a few SLE patients. As was discussed in the section on SSA/Ro and SS-B/La and demonstrated in Figure 4, it is common to detect antibodies to both SS-A/Ro 52 kD and Jo-1 in the same patient. Rutjes et.al reported that anti-SS-A/Ro 52 kD was detected in 20% of patients with idiopathic inflammatory myopathies but in patients with antibodies to Jo-1 the frequency was 58%. [79]. Antibodies to SS-A/Ro 52 were also detected in patients with antibodies to other synthetase proteins, SRP, and PM/Scl. [120]. In fact, antibodies to SS-A/Ro 52 are the most frequently detected antibodies in myositis.

Rheumatoid Arthritis

RA is the most common of the SRD. In 1800 a new form of polyarthritis that was different from gouty arthritis was described by Landré-Beauvais [121]. The term "rheumatoid arthritis" was first used to designate this form of polyarthritis in 1859. While many had thought that RA was a relatively new disease, that theory was challenged when skeletons dating back as far as 6500 years were found in the upper western Mississippi basin that displayed bilaterally symmetric joint erosions characteristic of RA.

The prevalence of RA is estimated to be about 0.5-1% in North American and European Caucasians over the age of 15 years. It is estimated that 20 men and 60 women per 100,000 will develop RA per year. The incidence increases with age.

RA is a clinical diagnosis. Patients commonly present to the clinician with a history of bilaterally symmetrical polyarthritis affecting at least 3 or more joints and morning stiffness lasting at least one hour. The joints most commonly affected are those of the hands, wrists, elbows, knees, ankles and feet. Radiographic changes occur in the joints indicating an erosive process. Between 50-70% of patients with erosions will continue to progress to tissue destruction. The majority of patients test positive for rheumatoid factor (RF). The disease is not limited only to joints. Extraarticular manifestations may include the vascular, cardiac, pulmonary, and renal systems. Pleurisy affects more than 70% of patients. Rheumatoid nodules occur in about a third of the patients. Secondary SS may develop in RF+ patients and occurs more frequently in females than males. It is differentiated from primary SS as patients lack antibodies to SS-A/Ro and SS-B/La.

RF has been the primary antibody associated with RA for over 60 years [122]. Antibodies to several other proteins have been described to be associated with RA [123]. To date, assays for the detection of most of these antibodies have not become commercially available. Antibodies considered highly specific but less sensitive for RA include antikeratin, antiperinuclear factor, and antifilaggrin. The common feature of these antibodies is that all react with proteins that are citrullinated. Citrullination occurs when the positively charged amino-group of arginine is hydrolyzed to a neutral oxygen-group [124]. A new ELISA, anti-cyclic citrullinated peptide (CCP), has been developed to detect these antibodies.

RFs are immunoglobulins of any isotype with antibody activity directed toward antigenic sites on the Fc region of human or animal IgG. They tend to bind to the Cγ2 or Cγ3 of IgG. Many RFs frequently require both domains for reactivity suggesting the presence of conformational epitopes. The RF response is primarily polyclonal. Most assays for RF detect antibodies to IgM RF. Tests for IgG and IgA RF are also available [122].

As mentioned previously, RF was first described by Waaler [12]. He observed that RA sera agglutinated sheep RBC coated with rabbit antibodies to sheep RBC. Later other particles, including latex, were substituted for the sheep RBC [125]. The Singer-Plotz latex agglutination assay using latex coated with aggregated human IgG and the Waaler-Rose assay became the standard methods for quantitation of RF. Newer assays, including nephelometry and ELISA, that allow for automation have replaced these standard agglutination assays in larger clinical laboratories.

RF is not specific for RA. It is detected in other autoimmune diseases including SS, SLE, MCTD, vasculitis, and idiopathic pulmonary fibrosis. It is not limited to autoimmune disease but may be detected in certain chronic infectious diseases as well as lymphoproliferative disorders such as Waldenström's macroglobulinemia, B-cell lymphoma, and chronic lymphocytic leukemia. In chronic lymphocytic leukemia the RF may be monoclonal rather than polyclonal as is generally the case.

Antibodies to perinuclear factor were first described in 1964 in patients with rheumatoid arthritis [126]. The test is an IIF assay where the patient's sera are reacted on buccal mucosal cell smears. The antibodies react with a number of keratohyaline granules surrounding the nucleus in selected cells. APF are reported to be present in 49-91% of RA patients with a specificity of 73-99% depending on the study. This test has not been readily available for routine use as only select individuals have mucosal cells that will work in the assay.

The presence of antibodies to keratin in RA patients was described in 1979 [127]. This assay is also IIF utilizing cryostat sections of rat esophagus. Anti-keratin antibodies stain the stratum corneum of rat epithelium in these sections. The sensitivity for RA varies from 36-59% and the specificity from 88-99%. This test is not readily available in routine labs.

Western blot studies using epidermal extracts demonstrated that many patients with RA react with a 40 kD protein identified as a neutral/acidic isoform of filaggrin [128]. Further studies demonstrated that a monoclonal antibody to (pro) filaggrin and antibodies to the perinuclear factor gave the same IFF stain. When affinity purified antibody from the 40 kD protein band was used in IIF testing, it produced characteristic staining patterns in both the perinuclear and keratin IIF assays suggesting a relationship [129].

Profilaggrin is post translationally modified in two ways. It is proteolytically cleaved during cellular differentiation to become filaggrin and ~20% of the arginine residues are converted to citrulline.

Filaggrin protein was analyzed to determine which sequences might be the antigenic epitopes for the RA antibodies. Selected peptides were synthesized and evaluated by ELISA for reactivity. Results demonstrated that sera from RA patients bound to certain of the citrullinated peptides [42].

Based on this study, a variant cyclic peptide containing citrulline was designed and used as antigenic substrate in ELISA [43]. The patients in this study included both RA and non-RA patients, patients with infectious disease, and patients from an early arthritis clinic. In comparing RA vs non-RA the specificity was 98% and the sensitivity was 68%. In the screening of patients from the early arthritis clinic the specificity was 96%. Comparing the standard IgM RF ELISA to the CCP ELISA, the specificity was 91% vs 96% and the sensitivity was 54% vs 48%. When the results were combined, the positive predictive value was 91%. They recommended that both IgM RF and CCP assays should be used to screen early arthritis patients. Because of the useful results from the anti-CCP ELISA, this test is now commercially available.

In order to determine the predictive value of anti-CCP antibodies in patients with early arthritis, 273 RA patients who had disease symptoms of <1 year were evaluated over 3-year and 6-year periods in terms of physical disability, radiologic damage, and the presence of anti-CCP. The results indicated that 70% of patients have anti-CCP early in the disease. Patients with anti-CCP developed significantly more severe radiologic damage than those without anti-CCP [130]. A second group tested 98 RA patients and 232 controls including age-matched healthy subjects for both RF and anti-CCP. They reported that anti-CCP alone was 41% sensitive and 97.8% specific while RF alone was 62% sensitive and 84% specific. With the combination of both assays, the specificity was 99.6%. They suggest, as did the previous study, that both tests be used for screening for early RA [131]. Thus, the development of an ELISA based assay for citrullinated proteins allows the detection of these antibodies to become more readily available, reproducible, and automated. An immunoblot blot assay using filaggrin enriched human epidermis extract has also been described [132].

Conclusion

It can be readily appreciated that certain autoantibodies, individually or in combination, are characteristic of a specific SRD. This is observed as antibodies to nDNA and Sm are rarely seen in any disease other than SLE, Scl-70 and centromere antibodies are associated with SSc, SS-A/Ro and SS-B/La antibodies together are characteristic of SS, RNP antibodies alone in high titer are seen in MCTD, Jo-1 and other t-RNA synthetase antibodies are diagnostic of PM, and CCP antibodies are detected in RA.

The relationships or linkage of antibody reactivity occurs as the proteins detected often are related to specific organelles or macromolecular complexes. For example, in SLE the major antibodies are directed to chromatin and the spliceosome, in SSc they are directed to nucleolar related proteins while in PM/DM they are directed to proteins in the cytoplasm associated with protein translation.

The autoantibodies of clinical significance in SRD are predominantly IgG and appear to result from an antigen driven response. They tend to react with antigenic sites that cover the entire surface of the molecule including sites that inhibit function. The epitopes with which they react may be linear but many are conformational. Some epitopes require two different proteins or a protein/nucleic acid complex to display reactivity. Thus, depending on the source, preparation, and presentation of the antigen, it is readily apparent why different results will be obtained between assays using the same serum.

While many patients with SRD can be readily diagnosed through clinical observations, others may not be as well defined. This may occur when patients are early in the disease or present with overlap symptoms. With close observations to the patterns and titers observed on a HEp-2 ANA screen and the application of specific antibody testing, the laboratory will aid in the diagnosis of these patients.

Bibliography

General References

1. *Textbook of the Autoimmune Diseases*. Lahita RG ed. Lippincott Williams & Wilkins, New York. 2000.

2. *Autoantibodies*. Peter JB and Y Shoenfeld ed. Elsevier, New York. 1996.

3. *Clinical Immunology: Principles and Practice 2nd ed*. Rich RR, TA Fleisher, WT Shearer, BL Kotzin and HW Schroeder Jr. ed. Vol I-II. Mosby, New York. 2001.

4. *Manual of Clinical Laboratory Immunology, 5th ed*. Rose NR, EC de Macario, JD Folds, HC Lane, RM Nakamura ed. ASM Press, Washington, D.C. 1997.

5. Crowle AJ. *Immunodiffusion 2nd ed*. Academic Press, New York. 1973.

6. Ouchterlony O *Handbook of Immunodiffusion and Immunoelectrophoresis*. Ann Arbor Science Publishers, Inc. Ann Arbor, MI. 1968.

7. *Laboratory Methods for the Detection of Antinuclear Antibodies*. Cavallaro JJ, FC McDuffie, MG Byrd, and JS McDougal ed. U.S. Dept. of HHS, CDC, Atlanta, GA. 1987

8. Nakamura RM, CL Peebles, RL Rubin, DP Molden and EM Tan. Autoantibodies to nuclear antigens (ANA). *Advances in laboratory tests and significance in systemic rheumatic diseases, 2nd ed*. ASCP, Chicago, Ill. 1984.

Review articles

9. Tan EM. Autoantibodies to Nuclear Antigens (ANA): Their Immunobiology and Medicine. In: *Advances in Immunology* 33: 167-240. 1982.

10. Tan EM. Antinuclear Antibodies: Diagnostic Markers for Autoimmune Diseases and Probes for Cell Biology. In: *Advances in Immunology* 44: 95-151. 1989.

11. Von Mühlen CA and EM Tan. Autoantibodies in the Diagnosis of Systemic Rheumatic Diseases. *Seminars in Arthritis and Rheumatism* 24: 323-358. 1995.

Selected References

12. Waaler E. On the occurrence of a factor in human serum activating the specific agglutination of sheep blood corpuscles. *Acta. Pathol. Microbiol. Scand.* 17:172-178. 1940.

13. Hargraves MM, H Richmond and R Moreton. Presentation of two bone marrow elements: The "tart" cells and the "L.E." cell. *Mayo Clin. Proc.* 27: 25-28. 1948.

14. Holman R and HR Deicher. The reaction of the lupus erythematosus (L.E.) cell factor with deoxyribonucleoprotein of the cell nucleus. *J. Clin. Invest.* 38: 2059-2072. 1959.

15. Tan EM. The LE cell and antinuclear antibodies: a breakthrough in diagnosis. In: *Landmark Advances in Rheumatology,* Faucher GJ, ed. Contact Associates International Ltd. 1985.

16. Coons AH, HJ Creech , RN Jones and E Berliner. The demonstration of pneumococcal antigen in tissues by the use of the fluorescent antibody. *J. Immunol* 45: 159-170. 1942.

17. Coons AH and MH Kaplan. Localization of antigen in tissue cells: II. Improvements in a method for detection of antigen by means of fluorescent antibody. *J Exper. Med.* 91: 1-13. 1950.

18. Weller TH and AH Coons. Fluorescent antibody studies with agents of varicella and herpes zoster propagated *in vitro. Proc. Soc. Exp. Biol. (N.Y.)* 86: 789-794. 1954.

19. Holman H and HG Kunkel. Affinity between the lupus erythematosus serum factor and cell nuclei and nucleoprotein. *Science* 126: 162. 1957.

20. Friou GJ. Clinical application of lupus serum—nucleoprotein reaction using the fluorescent antibody technique. (Abstr.) *Jour. Clin. Invest.* 36: 890. 1957.

21. Holborow EJ, DM Weir and GD Johnson. A serum factor in lupus erythematosus with affinity for tissue nuclei. *Brit. Med. J.* 5047: 732-734. 1957.

22. Beck JS. Antinuclear antibodies: methods of detection and significance. *Mayo Clin. Proc.* 44: 600-619. 1969.

23. Diecher H, H Holman and H Kunkel. The Precipitin between DNA and a Serum Factor in Systemic Lupus Erythematosus. *J. Exp. Med.* 109: 97-114. 1959.

24. Tan EM, RI Carr, PH Schur and HG Kunkel. DNA and antibody to DNA in the serum of patients with systemic lupus erythematosus. *J. Clin. Invest.* 45: 1732-1740. 1966.

25. Koffler D, PH Schur, and HG Kunkel. Immunological studies concerning the nephrites of systemic lupus erythematosus. *J. Exp. Med.* 126: 607-634. 1967.

26. Wold RT, FE Young, EM Tan and RS Farr. Deoxyribonucleic acid antibody: a method to detect its primary interaction with deoxyribonucleic acid. *Science:* 161: 806-807. 1968.

27. Anderson JR, KG Gray, JS Beck, WW Buchanan, and AJ McElhinney. Precipitating auto-antibodies in the connective tissue diseases. *Ann. Rheum. Dis.* 21: 360-369. 1962.

28. Beck JS, JR Anderson, KG Gray, and NR Rowell. Antinuclear and precipitating autoantibodies in progressive systemic sclerosis. *The Lancet:* Dec. 7: 1188-1190. 1963.

29. Tan EM and HG Kunkel. Characteristics of a soluble nuclear antigen precipitating with sera of patients with systemic lupus erythematosus. *J. Immunol.* 96: 464-471. 1966.

30. Northway JD and EM Tan. Differentiation of antinuclear antibodies giving speckled staining patterns in immunofluorescence. *Clin. Immunol. Immunopathol.* 1: 140-152. 1972.

31. Notman DD, N Kurata and EM Tan. Profiles of antinuclear antibodies in systemic rheumatic disease. *Ann. Intern. Med.* 83: 464-469. 1975.

32. Kurata N and EM Tan. Identification of antibodies to nuclear acidic antigens by counterimmunoelectrophoresis. *Arthritis Rheum.* 19: 574-580. 1976.

33. Sharp GC, WS Irvin, RL Laroque, C Velez, V Daly, AD Kaiser, and HR Holman. Mixed connective tissue disease—an apparently distinct rheumatic disease syndrome associated with a specific antibody to an extractable nuclear antigen (ENA). *Am. J. Med.* 52: 148-159. 1972.

34. Clarke G, M. Reichlin and TB Tomasi. Characterization of a soluble cytoplasmic antigen reactive with sera from patients with systemic lupus erythematosus. *J. Immunol.* 102: 117-124. 1969.

35. Mattioli M and M Reichlin. Heterogeneity of RNA protein antigens reactive with sera of patients with systemic lupus erythematosus. *Arthritis Rheum.* 17: 421-429. 1974.

36. Alspaugh MA D. Talal, and EM Tan. Differentiation and characterization of autoantibodies and their antigens in Sjögren's syndrome. *Arthritis Rheum.* 19: 216-222. 1976.

37. Alspaugh MA and P Madison. Resolution of the identity of certain antigen-antibody systems in systemic lupus erythematosus and Sjögren's syndrome: An interlaboratory collaboration. *Arthritis Rheum.* 22: 796. 1979.

38. Douvas AS, M Achten and EM Tan. Identification of a nuclear protein (Scl-70) as a unique target of human antinuclear antibodies in scleroderma. *J. Biol. Chem.* 254: 10514-10522. 1979.

39. Bernstein RM, JC Steigerwald, and EM Tan. Association of antinuclear and antinucleolar antibodies in progressive systemic sclerosis. *Clin. Exp. Immunol.* 48: 43-51. 1982.

40. Moroi Y, C Peebles, MJ Fritzler, J Steigerwald and EM Tan. Autoantibody to centromere (kinetochore) in scleroderma sera. *Proc. Natl. Acad. Sci.* 77: 1628-1631. 1980.

41. Nishikai M and M Reichlin. Heterogeneity of precipitating antibodies in polymyositis: Characterization of the Jo-1 antibody system. *Arthritis Rheum.* 23: 881-888. 1980.

42. Shellekens GA, BAW de Jong, FHJ van den Hoogen, LBA van de Putte, and WJ van Venrooij. Citrulline is an essential constituent of antigenic determinants recognized by rheumatoid arthritis-specific autoantibodies. *J. Clin. Invest.* 101: 273-281. 1998.

43. Shellekens GA , H Visser, BA de Jong, FH van den Hoogen, JM Hazes, FC Breedveld, and WJ Venrooij. The diagnostic properties of rheumatoid arthritis antibodies recognizing a cyclic citrullinated peptide. *Arthritis Rheum.* 43: 155-163. 2000.

44. Aarden LA, ER deGroot, and TEW Feltkamp. Immunology of DNA. III. Crithidia luciliae, a simple substrate for the determinatin of anti-dsDNA with the immunofluorescence technique. *Ann. N.Y. Acad. Sci.* 254: 505-515. 1975.

45. Rubin RL. Enzyme-linked immunosorbent assays for antibodies to native DNA, histones, and (H2A-H2B)-DNA. In: *Manual of Clinical Laboratory Immunology,* 5th ed. Rosa NR, EC de Macario, JD Folds, HC Lane, RM Nakamura ed. ASM Press, Washington, D.C. 1997. pp 935-938.

46. Nakamura RM, CL Peebles, and EM Tan. Microhemaggllutination test for detection of antibodies to nuclear Sm and Ribonucleoprotein antigens in systemic lupus erythematosus and related diseases. *Amer. J. Clin. Path.* 70: 800-807. 1978.

47. Tan EM and C Peebles. Quantitation of antibodies to Sm antigen and nuclear ribonucleoprotein by hemagglutination. In: *Manual of Clinical Immunology,* 2nd ed. Rose NR and H Friedman ed. ASM Press, Washington, D.C. pp 866-870. 1980.

48. Hollingsworth PN, SC Pummer and RL Dawkins. Antinuclear antibodies. In: *Autoantibodies* Peter JB and Y Schoenfeld eds. Elsiever Science N.Y. 1996. pp 74-90.

49. Humbel RL. Detection of antinuclear antibodies by immunofluorescence. In: *Manual of Biological Markers of Disease.* Kluwer Academic Publisher. The Netherlands. 1993. ppA2:1-16.

50. Kavanaugh A, R Tomar, J Reveille, DH Solomon and HA Homberger. Guidelines for clinical use of the antinuclear antibody test and tests for specific autoantibodies to nuclear antigens. *Arch. Path. & Lab. Med.* 124:71-81. 2000.

51. Lahita RG. Systemic lupus erythematosus. In: *Textbook of the Autoimmune Diseases*. Lahita RG ed. Lippincott Williams & Wilkins, New York. 2000. pp 537-547.

52. Harmon CE, JS Deng, CL Peebles, and EM Tan. The importance of tissue substrate in the SS-A/Ro antigen-antibody system. *Arthritis Rheum.* 27: 166-173. 1984.

53. Amoura Z, JC Piette, JF Bach, and S Koutouzov. The key role of nucleosomes in lupus. *Arth. Rheum.* 42: 833-843. 1999.

54. Winfield JB, D Koffler and HG Kunkel. Role of DNA-anti-DNA complexes in the immunopathogenesis of tissue injury in systemic lupus erythematosus. *Scand. J. Rheumatol.,* suppl. 11: 59-64. 1975.

55. Winfield JB, I Faiferman, and D Koffler. Avidity of anti-DNA antibodies in serum and IgG glomerular eluates from patients with systemic lupus erythematosus. *J. Clin. Invest.* 59: 90-96. 1977.

56. Hill GS, N Hinglais, F Tron, and JF Bach. Systemic lupus erythematosus, morphologic correlations with Immunologic and clinical data at the time of biopsy. *Am. J. Med.* 64: 61-79. 1978.

57. Smeenk RJT, JHM Berden, AJG Swaak. DsDNA Autoantibodies. In: *Autoantibodies*. Peter JB and Y Shoenfeld ed. Elsevier, New York. 1996. pp 227-236.

58. Rubin RL, FG Joslin, EM Tan. An improved ELISA for anti-native DNA by elimination of interference by anti-histone antibodies. *J. Immunol. Methods* 63: 359-366. 1983.

59. Tan EM, JS Smolen, JS McDougal, BT Butcher, D Conn, R Dawkins, MJ Fritzler, T Gordon, JA Hardin, JR Kalden, RG Lahita, RN Maini, NF Rothfield, R Smeek, Y Takasaki, WJ van Venrooij, A Wiik, M Wilson, and JA Koziol. A critical evaluation of enzyme immunoassays for detection of antinuclear autoantibodies of defined specificities. I. Precision, Sensitivity, and Specificity. *Arthritis Rheum.* 42: 455-464. 1999.

60. McGuigan L, J Edmonds, G Wellings, and M Jones. The significance of discrepant Farr and *Crithidia luciliae* tests. *J. Rheum.* 11: 172-174. 1984.

61. Burlingame RW, RL Rubin, RS Balderas and AN Theofilopoulos. Genesis and evolution of antichromatin autoantibodies in murine lupus implicates T-dependent immunization with self antigen. *J. Clin. Invest.* 91: 1687-1696. 1993.

62. Burlingame RW, ML Boey, G Starkebaum, and RL Rubin. The central role of chromatin in autoimmune responses to histones and DNA in systemic lupus erythematosus. *J. Clin. Invest.* 94: 184-192. 1994.

63. Burlingame RW and RL Rubin. Subnucleosome structures as substrates in enzyme-linked immunosorbent assays. *J. Immunol. Methods* 134: 187-199. 1990.

64. Nakamura RM, CL Peebles, RL Rubin, DP Molden and EM Tan. Autoantibodies to nuclear antigens (ANA). Counterimmunoelectrophoresis. In: *Advances in laboratory tests and significance in systemic rheumatic diseases, 2nd ed.* ASCP, Chicago, Ill. 1984. pp 116-121.

65. Lerner MR and JA Steitz. Antibodies to small nuclear RNAs complexed with proteins are produced by patients with systemic lupus erythematosus. *Proc. Natl. Acad. Sci. USA* 76: 5495-5499. 1979.

66. Chan EKL, C Peebles, MJ Fritzler, M Satoh. Anti-Spliceosomal Autoantibodies. IN: *Systemic lupus erythematosus, 1st ed.* Toskos GC, C Gordon, JS Smolen ed. Mosby-Elsevier, Philadelphia, PA. 2007. pp274-280.

67. Tan EM, MJ Fritzler, JS McDougal, FC McDuffie, RM Nakamura, M Reichlin, CB Reimer, GC Sharp, PH Schur, MR Wilson, RJ Winchester. Reference sera for antinuclear antibodies. I. Antibodies to native DNA, Sm, nuclear RNP, and SS-B/La. *Arthritis Rheum.* 25: 1003-1005. 1982.

68. Smolen Js, B Butcher, MJ Fritzler, T Gordon, J Hardin, JR Kalden, R Lahita, RN Maini, W Reeves, M Reichlin, N Rothfield, Y Takasaki, WJ van Venrooij, and EM Tan. Reference sera for antinuclear antibodies. II. Further definition of antibody specificities in international antinuclear antibody reference sera by immunofluorescence and western blotting. *Arthritis Rheum.* 40: 413-418. 1997.

69. Franco HL, WL Weston, C Peebles, SL Forstot, and P Phanuphak. Autoantibodies directed against sicca syndrome antigens in neonatal lupus syndrome. *J. Amer. Acad. Dermatol.* 4: 67-72. 1981.

70. Buyon JP, E Ben-Chetrit, S. Karp, RA Roubey, L Pompeo, WH Reeves, EM Tan, and R Winchester. Acquired congenital heart block. Pattern of maternal antibody response to biochemically defined antigens of the SS-A/Ro--SS-B/La system in neonatal lupus. *J. Clin. Invest.* 84: 627-634. 1989.

71. Alexander EL, JP Buyon, J Lane, A Lafond-Walker, TT Provost, T Guarnieri. Anti-SS-A/Ro SS-B/La antibodies bind to neonatal rabbit cardiac cells and preferentially inhibit in vitro cardiac repolarization. *J. Autoimmun.* 2: 463-469. 1989.

72. Chan EKL and LEC Andrade. Antinuclear antibodies in Sjögren's syndrome. *Rheumatic Disease Clinics of North America* 18: 551-570. 1992.

73. Fabini G, SA Rutjes, C Zimmermann, GJM Pruijn, and G Steiner. Analysis of the molecular composition of Ro ribonucleoprotein complexes: Identification of novel Y RNA-binding proteins. *Eur. J. Biochem.* 267: 2778-2789. 2000.

74. Reymond A, G Meroni, A Fantozzi, G Meria, S Cairo, L Luzi, D Riganelli, E Zanaria, S Messali, S Cainarca, A Gurranti, S Minucci, PG Pelicci, and A Ballabio. The tripartite motif family identifies cell compartments. *EMBO* 20:2140-2151. 2001.

75. Wada K, T Kamitani. Autoantigen Ro52 is an E3 ubiquitin ligase. *Biochem Biophys Res Commun.* 339: 415-421. 2006

76. A Espinoza, W Zhou, M Ek, M Hedlund, S Brauner, K Popovic, L Horvath, T Wallerskog, M. Oukka, F Nyberg, VK Kuchroo andM Wahren-Herlenius. The Sjögren's Syndrome-Associated Autoantigen Ro52 is an E3 Ligase That Regulates Proliferation and Cell Death. *J Imm* 176: 6277-6285. 2006.

77. Ben-Chetrit E, EKL Chan, KF Sullivan, and EM Tan. A 52-kD Protein is a Novel Component of the SS-A/Ro Antigenic Particle. *J. Exp Med* 167: 1560-1571. 1988.

78. Ben-Chetrit E, RI Fox and EM Tan. Dissociation of the immune responses to the SS-A (Ro) 52-kd and 60-kd polypeptides in systemic lupus erythematosus and Sjögren's syndrome. *Arthritis Rheum.* 33: 349-355. 1990.

79. Rutjes SA, WT Vree Egberts, P Jongen, F Van Den Hoogan, GJ Pruijn, and WJ Van Venrooij. Anti-Ro52 antibodies frequently co-occur with anti-Jo-1 antibodies in sera from patients with idiopathic inflammatory myopathy. *Clin. Exp. Immunol.* 109: 32-40. 1997.

80. Martin AL and M Reichlin. Fluctuations of antibody to ribosomal P proteins correlate with appearance and remission of nephritis is SLE. *Lupus:* 5: 22-29. 1996

81. Chindalore V, B Neas, and M Reichlin. The association between anti-ribosomal P antibodies and active nephritis in systemic lupus erythematosus. *Clin. Immunol. Immunopathol.* 87: 292-296. 1998.

82. Francoeur AM, CL Peebles, KJ Heckman, JC Lee, and EM Tan. Identification of ribosomal protein autoantigens. *J. Immunol.* 135: 2378-2384. 1985.

83. Dwyer E and RG Lahita. Ribosomal autoantibodies. In: *Autoantibodies*. Peter JB and Y Shoenfeld ed. Elsevier, New York. 1996. pp 716-720.

84. Miyachi K, MJ Fritzler and EM Tan. Autoantibody to a nuclear antigen in proliferating cells. *J. Immunol.* 121: 2228-2234. 1978.

85. Takasaki Y, JS Deng and EM Tan. A nuclear antigen associated with cell proliferation and blast transformation. *J. Exp. Med.* 1154: 1899-1909. 1981.

86. Lahita RG. Sjögren's Syndrome. In: *Textbook of the Autoimmune Diseases.* Lahita RG Ed. Lippincott Williams & Wilkins, New York. 2000. pp 569-572.

87. Fox RI. Sjögren's Syndrome. In: *Clinical Immunology: Principles and Practice 2nd ed.* Rich RR, TA Fleisher, WT Shearer, BL Kotzin and HW Schroeder Jr. ed. Vol I. Mosby, New York. 2001. pp 63.1-11.

88. Smith MD. Scleroderma. In: *Textbook of the Autoimmune Diseases*. Lahita RG Ed. Lippincott Williams & Wilkins, New York. 2000. pp 557-567.

89. Tan EM, GP Rodnan, I Garcia, Y Moroi, MJ Fritzler, C Peebles. Diversity of antinuclear antibodies in progressive systemic sclerosis. Anti-centromere antibody and its relationship to CREST syndrome. *Arthritis Rheum.* 23: 617-625. 1980.

90. Guldner HH, C, Szosteki, HP Vosberg, HJ Lakomek, E Penner, and FA Bautz. Scl-70 autoantibodies from scleroderma patients recognize a 95 kDa protein identified as topoisomerase I. *Chromosoma* 94: 132-138. 1986.

91. Shero JH, B Bordwell, NF Rothfield, and WC Earnshaw. High titiers of autoantibodies to topoisomerase I (Scl-70) in sera from scleroderma patients. *Science* 231: 737-740. 1986.

92. Maul GG, BT French, WJ van Venrooij, and SA Jimenez. Topoisomerase I identified by scleroderma 70 antisera: enrichment of topoisomerase I at the centromere in mouse mitotic cells before anaphase. *Proc. Natl. Acad. Sci. USA* 83: 5145-5149. 1986.

93. Jarzabek-Chorzelska M, M Blaszczyk, S Jablonska, T Chorzelski, V Kumar, and EH Beutner. Scl-70 antibody—a specific marker of systemic sclerosis. *Br. J. Dermatol.* 115: 393-401. 1986.

94. Earnshaw WC, BJ Bordwell, C Marino, and NF Rothfield. Three human chromosomal autoantigens are recognized by sera from patients with anti-centromere antibodies. *J. Clin. Invest.* 77: 426-430. 1986.

95. Reimer G, KM Rose, U Scheer and EM Tan. Autoantibody to RNA polymerase I in scleroderma sera. *J. Clin. Invest.* 79: 65-72. 1987.

96. Targoff IN, and M Reichlin. Nucleolar localization of the PM-Scl antigen. *Arthritis Rheum.* 28: 226-230. 1985.

97. Reimer G, U Scheer, JM Peters, and EM Tan. Immunolocalization and partial characterization of a nucleolar autoantigen (PM-Scl) associated with polymyositis/scleroderma overlap syndromes. *J. Immunol.* 137: 3802-3808. 1986.

98. Hashimoto C, and JA Steitz. Sequential association of nucleolar 7-2 RNA with two different autoantigens. *J. Biol. Chem* 258: 1379-1382. 1983.

99. Reddy R, EM Tan, D Henning, K Nohga, and H Busch. Detection of a nucleolar 7-2 ribonucleoprotein and a cytoplasmic 8-2 ribonucleoprotein with autoantibodies from patients with scleroderma. *J. Biol. Chem* 258: 1383-1386. 1983.

100. Reimer G, I Raska, U Scheer, EM Tan. Immunolocalization of 7-2-ribonucleoprotein in the granular component of the nucleolus. *Exp. Cell Res.* 176: 117-128. 1988.

101. Ochs RL, MA Lischwe, WH Spohn, and H Busch. Fibrillarin: a new protein of the nucleolus identified by autoimmune sera. *Biol. Cell* 54: 123-133. 1985.

102. Reimer G, KM Pollard, CA Penning, RL Ochs, MA Lischwe, H Busch and EM Tan. Monoclonal autoantibody from NZB/NZW F1 mouse and some human scleroderma sera target a Mr 34,000 nucleolar protein of the U3-ribonucleoprotein particle. *Arthritis Rheum.* 30: 793-800. 1987.

103. Couzin J. Stripping the nucleolus down to its proteins. *Science* 295: 422. 2002.

104. Arnett FC, JD Reveille, R Goldstein, KM Pollard, K Leaird, EA Smith, EC Leroy, and MJ Fritzler. Autoantibodies to fibrillarin in systemic sclerosis (scleroderma). An immunogenetic, serologic and clinical analysis. *Arthritis Rheum.* 39: 1151-1160. 1996.

105. Tormey VJ, CC Bunn, CP Denton, and CM Black. Anti-fibrillarin antibodies in systemic sclerosis. *Rheumatology (Oxford)* 40: 1157-1162. 2001.

106. Reveille JD, M Fischbach, T McNearney, AW Friedman, MB Aguilar, J Lisse, MJ Fritzler, C Ahn and FC Arnett. Systemic sclerosis in 3 US ethnic groups: a comparison of clinical, socio-demographic, serologic and immunogenetic determinants. *Semin. Arthritis Rheum.* 30: 332-346. 2001.

107. Okano Y and TA Medsger Jr. Autoantibody to Th ribonucleoprotein (nucleolar 7-2 RNA protein particle) in patients with systemic sclerosis. *Arthritis Rheum.* 33: 1822-1828. 1990.

108. Yamane K, H Ihn, M Kubo, M Kuwana, Y Asano, N Yazawa, and K Tamaki. Antibodies to Th/To ribonucleoprotein in patients with localized scleroderma. *Rheumatology (Oxford)* 40: 683-686. 2001.

109. Wolfe JF, E Adelstein, and GC Sharp. Antinuclear antibody with distinct specificity for polymyositis. *J. Clin. Invest.* 59: 176-178. 1977.

110. Reichlin M, PJ Maddison, I Targoff, T Bunch, F Arnett, G Sharp, E Treadwell, and EM Tan. Antibodies to a nuclear/nucleolar antigen in patients with polymyositis overlap syndromes. *J. Clin. Immunol.* 4: 40-44. 1984.

111. Alderuccio F, EK Chan, and EM Tan. Molecular characterization of an autoantigen of PM-Scl in the polymyositis/scleroderma overlap syndrome: a unique and complete human c DNA encoding an apparent 75-kD acidic protein of the nucleolar complex. *J. Exp. Med.* 173: 941-952. 1991.

112. Ge Q, MB Frank, C O'Brien, IN Targoff. Cloning of a complementary DNA coding for the 100-kD antigenic protein of the PM-Scl autoantigen. *J. Clin. Invest.* 90: 559-570. 1992.

113. Kuwana M, Y Okana, J Kaburaki, T Tojo, TA Medsger, Jr. Racial differences in the distribution of systemic sclerosis-related serum antinuclear antibodies. *Arthritis Rheum.* 37: 902-906. 1994.

114. Okano Y and TA Medsger, Jr. RNA polymerase I-III autoantibodies. In: *Autoantibodies*. Peter JB and Y Shoenfeld ed. Elsevier, New York. 1996. pp 727-734.

115. Kuwana M, J Kaburaki, T Mimori, T Tojo, M Homma. Autoantibody reactive with three classes of RNA polymerases in sera from patients with systemic sclerosis. *J. Clin. Invest.* 91: 1399-1404. 1993.

116. Kuwana M, K Kimura and Y Kawakami. Identification of an Immunodominant Epitope on RNA Polymerase III recognized by Systemic Sclerosis Sera. *Arthritis Rheum* 46: 2742-2747. 2002.

117. Kuwana M, Y Okano, JP Pandey, RM Silver, N Fertig, and TA Medsger, Jr. Enzyme-linked Immunosorbent Assay for Detection of Anit-RNA Polymerase III Antibody. *Arthritis Rheum* 52: 2425-2432. 2005.

118. Rider LG and IN Targoff. Muscle diseases. In: *Textbook of the Autoimmune Diseases*. Lahita RG Ed. Lippincott Williams & Wilkins, New York. 2000. pp 429-474.

119. Rosa MD, JP Hendrick jr, MR Lerner, JA Steitz, and M Reichlin. A mammalian tRNA His-containing antigen is recognized by the polymyositis-specific antibody anti-Jo-1. *Nucleic Acids Res.* 11: 853-870. 1983.

120. Frank MB, V McCubbin, E Triieu, Y Wu, DA Isenberg, and IN Targoff. The association of the anti-Ro52 autoantibodies with myositis and scleroderma autoantibodies. *J. Autoimmun.* 12: 137-142. 1999.

121. Weyand CM and JJ Goronzy. Rheumatoid arthritis. In: *Textbook of the Autoimmune Diseases*. Lahita RG Ed. Lippincott Williams & Wilkins, New York. 2000. pp 573-594.

122. Wener MH and M Mannik. Rheumatoid factors. In: *Manual of Clinical Laboratory Immunology, 5th ed.* Rose NR, EC de Macario, JD Folds, HC Lane, RM Nakamura ed. ASM Press, Washington, D.C. 1997. pp 942-948.

123. Goldbach-Mansky R, J Lee, A McCoy, J Hoxworth, C Yarboro, JS Smolen, G Steiner, A Rosen, C Zhang, HA Ménard, ZJ Zhou, T Palosuo, WJ Van Venrooij, RL Wilder, JH Klippel, HR Schumacher Jr, and HS El-Gabalawy. Rheumatoid arthritis associated autoantibodies in patients with synovitis of recent onset. *Arthritis Res.* 2: 236-243. 2000.

124. van Venrooij WJ and GJM Pruijn. Citrullination: a small change for a protein with a great consequences for rheumatoid arthritis. *Arthritis Res.* 2: 249-251. 2000.

125. Singer JM and CM Plotz. The latex fixation test. I. Application to the serological diagnosis of rheumatoid arthritis. *Am. J. Med.* 21: 888-892. 1956.

126. Nienhuis RLF and EA Mandema. A new serum factor in patients with rheumatoid arthritis. The antiperinuclear factor. *Ann. Rheum. Dis.* 23: 302-305. 1964.

127. Young BJJ, RK Mallya, RDJ Leslie, CJM Clark, and TJ Hamblin. Anti-keratin antibodies in rheumatoid arthritis. *Br. Med. J.* 2: 97-99. 1979.

128. Simon M, E Girbal, M Sebbag, V Gomés-Daudrix, C Vincent, G Salema, and G Serre. The cytokeratin filament-aggregating protein filaggrin is the target of the so-called "antikeratin antibodies" autoantibodies specific for rheumatoid arthritis. *J. Clin. Invest.* 92: 1387-1393. 1993.

129. Sebbag M, M Simon C Vincent, C Masson-Bessiere, E Girbal, JJ Derieux, and G Serre. The antiperinuclear factor and the so-called anti-keratin antibodies are the same rheumatoid arthritis-specific autoantibodies. *J. Clin. Invest.* 95: 2672-2679. 1995.

130. Kroot EJ, BA de Jong, MA van Leeuwen, H Swinkels, FH van den Hoogen, M van't Hof, LB van de Putte, MH van Ritjswijk, WJ van Venrooij, and PL van Riel. The prognostic value of anti-cyclic citrullinated peptide antibody in patients with recent-onset rheumatoid arthritis. *Arthritis Rheum.* 43: 1831-1835. 2000.

131. Bizzaaro N, G Mazzanti, E Tonutti, D Villalta, and R Tozzoli. Diagnostic accuracy of the anti-citrulline antibody assay for rheumatoid arthritis. *Clin. Chem.* 47: 1089-1093. 2001.

132. Vincent C, M Simon, M Sebbag, E Girbal-Neuhauser, JJ Durieux, A Cantagrel, B Fournie, B Mazieres, G Serre. Immunoblotting detection of autoantibodies to human epidermis filaggrin: a new diagnostic test for rheumatoid arthritis. *J. Rheumatol.* 25: 838-846. 1998.

Chapter 3

Laboratory Methodologies

Immunodiffusion

All of the major antibodies to the extractable nuclear antigens (ENA) in the systemic rheumatic diseases were originally detected and defined using Ouchterlony double immunodiffusion. Detailed information on the procedure may be found in books by Örjan Ouchterlony [1] and Alfred J Crowle [2]. In this type of double immunodiffusion both the antigen and the antibody are allowed to diffuse through a gel matrix where a positive reaction results in insoluble precipitates. This precipitation occurs at the point where the lattice formation caused by crosslinking of antigen and antibody becomes sufficiently large so as to become insoluble. The formation of this precipitin line is dependent on three major variables. These variables include the rate of diffusion, the relative concentrations of the reactants, and how soon they can form visible complexes after their fronts have met.

There are certain minimal criteria that must be met before precipitation can occur. The antibodies must be at least divalent with combining sites able to bind to separate antigens. The antigens must be multivalent with preferably more than one antigenic site available for crosslinking. The more sites/antigen, the better and more quickly it will precipitate.

Only two major reagents are needed for the identification of antibodies to the ENAs using immunodiffusion. The first is a control antibody of known specificity for the one to be identified. This is normally supplied in the test kits. It is good to have in house controls for comparison of results between lots of kits. To obtain an in house control set, collect and store extra serum on patients with defined specificity. This collection of sera allows not only for lot- to-lot comparisons but can be used for teaching new technologists or for testing new assays as they become available. A set of defined ANA specificities can be obtained from CDC [3]. The second reagent is the antigen. This is a cellular extract containing numerous antigens. The source of the extract and any pretreatment will affect the results. For instance, rabbit thymus, a common antigen source, will allow for the detection of antibodies to Sm, RNP, SS-B/La and Scl-70 but will not allow for the detection of antibodies to SS-A/Ro. SS-A/Ro can be detected using human tissue culture cell extracts or extracts of bovine thymus [4]. Extracts should be kept cold until use. Excessive heat will destroy the RNP antigen. Other antigens are labile and will degrade over time. Avoid freeze-thaw of antigens.

The procedure to compare antibodies is simple. Patient sera, control sera, and cellular extract are placed in adjacent wells in agarose gel plates. The samples are allowed to diffuse through the gel until precipitin lines are observed. This may be 24-72 hours depending on the system. As the extract contains numerous antigens and the sera may contain multiple antibodies, more than one line of reaction may occur. A number of conditions including the composition of the gel matrix, the pH of the buffers, the temperature of the reaction, the nature and concentration of the reactants and the size and distance between the wells affect the reactions. A discussion of all of these variables is beyond the scope of this paper. It is, however, important to the interpretation of the reactions to understand the effect of the antigen/antibody concentration on the presence and position of the individual precipitin lines. If the reactants are roughly equivalent, the precipitation bands will occur approximately equidistance from the reactant wells. If one of the reactants is greater in concentration, the band will shift toward the reactant of lower concentration. If one of the reactants is in gross excess, a reaction known as prozone may occur. Due to the excess reactant, only soluble complexes are formed and no lattice formation leading to visible precipitation can occur. In the case of ENA testing, it is usually the result of antibody excess. To optimize the conditions simply dilute the serum. Use the results of the FANA to help determine what may be an optimum dilution. By the use of specific prototypes and adjustment of the concentration of reactants, identification of many antibody specificities is possible. The lines observed are usually drawn on a template for a permanent record and results reported as positive or negative.

Originally, the most common use of Oüchterlony immunodiffusion was for the comparison of antigens. In the interpretation of the results three major types of reactions, identity, non-identity and partial identity were described to indicate the relationship between the antigens being tested. When applying this technique to the identification of autoantibodies, only identity and non-identity occur. (See Figure 1).

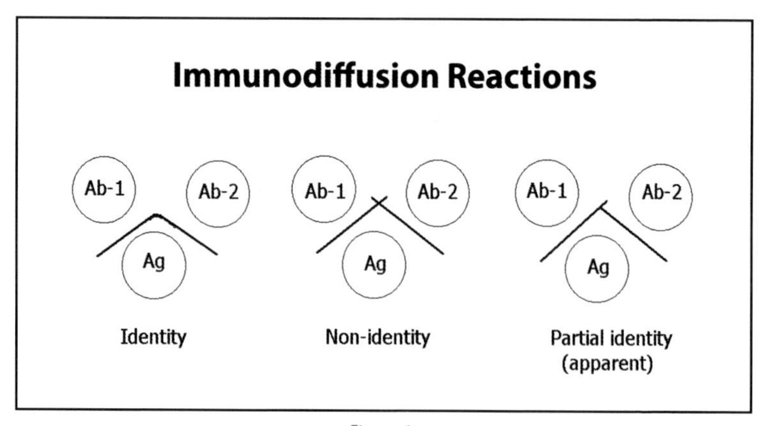

Figure 1.

The first of these, identity or fusion, results when the precipitin lines from both wells merge forming an arc. This indicates that the antibodies are reacting with the same antigen or complex of antigens. The second type, nonidentity, results when the lines cross forming an X indicating the antibodies present react with different antigens in the extract. In some instances one line may appear to merge into but not cross the other line creating what is described as a spur. This is seen when patients with antibodies to Sm are placed adjacent to patients with antibodies to RNP. (See Figure 2) This was erroneously described as "partial identity" which cannot occur with antibody comparison. The explanation of this spur formation was not possible until application

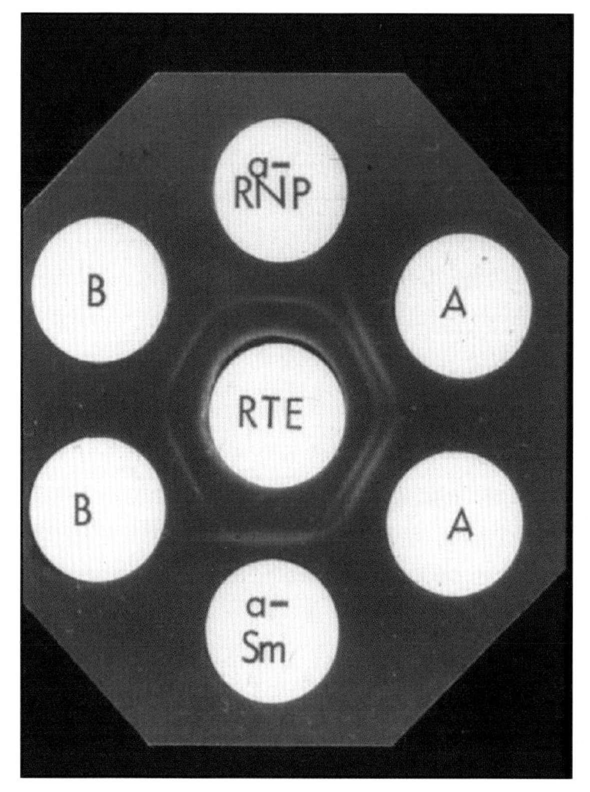

Figure 2. Immunodiffusion plate for the detection and identification of antibodies to Sm and RNP. The serum from patient A contains antibodies to both Sm and RNP while the serum from patient B contains antibodies to RNP only. Note the spur formed between the well containing antibodies to Sm and patient B. [5]

of the techniques of immunoblotting and immunoprecipitation. Sm and RNP antibodies recognize different proteins associated with a macromolecular complex known as the spliceosome. Sm antibodies react with major proteins in the complex designated B'/B and D. RNP antibodies react with major proteins in this complex designated 70 kD, A and C [6]. The extract used in immunodiffusion contains the Sm/ RNP complex containing all of the specific proteins as well as some free SmD protein. Because of this, antibodies to both RNP and Sm will react with the Sm/RNP complex. This results in the major reaction that appears identical. The spur develops because there is free SmD in the extract that can continue to diffuse past the precipitin reactivity to the complex and react with antibodies in the anti-Sm serum that are specific for the SmD protein. This is not a true partial identity but represents reactivity to a separate but related protein reacting in a location not allowing visible differences between the lines. There are technical conditions that may cause a spur to form one of which is a difference in the concentrations of the antibodies being compared. Care must always be taken in interpreting these results.

Immunodiffusion is highly specific but not very sensitive. A major advantage is that it requires no special equipment. The disadvantages are that, even though simple, it becomes labor intensive when running large numbers of assays and It takes from 24-72 hours to obtain results depending on the antibodies present.

1. Ouchterlony O. Handbook of Immunodiffusion and Immunoelectrophoresis. Ann Arbor Science Publishers, Inc. Ann Arbor, MI. 1968.

2. Crowle AJ. Immunodiffusion 2nd ed. Academic Press, New York. 1973.

3. *Laboratory Methods for the Detection of Antinuclear Antibodies.* Cavallaro JJ, FC McDuffie, MG Byrd, and JS McDougal ed. U.S. Dept. of HHS, CDC, Atlanta, GA. 1987.

4. Harmon CE, JS Deng, CL Peebles, and EM Tan. The importance of tissue substrate in the SS-A/Ro antigen-antibody system. *Arthritis Rheum.* 27: 166-173. 1984.

5. Nakamura RM, CL Peebles, RL Rubin, DP Molden, and EM Tan. Sm and U1-RNP autoantibodies: immunodiffusion assay. In: *Autoantibodies to nuclear antigens (ANA). Advances in laboratory tests and significance in systemic rheumatic diseases, 2nd ed.* ASCP Press, Chicago, Ill. 1984. pp 109-115.

6. Tan EM. Autoantibodies to Nuclear Antigens (ANA): Their Immunobiology and Medicine. In: *Advances in Immunology* 33: 167-240. 1982.

Counterimmunoelectrophoresis (CIE)

CIE is nothing more than an immunodiffusion reaction that has been accelerated by being placed in an electric field. Parallel wells are cut into a gel matrix precoated on a glass slide. Patient sera and control sera are added to the wells nearest the anode. Cellular extract is placed in the cathodic wells. The slide is placed in an electrophoresis chamber and a current applied for a specified period of time. This is usually 30-60 minutes. The lines observed are recorded on template and reported as positive or negative. Depending on the arrangement of the wells, identification is also possible.

The method relies on the charge of the proteins involved. The antibodies migrate toward the cathode while the antigens to be identified migrate toward the anode. When they cross over a precipitate forms creating a line. Reactions can be observed immediately after electrophoresis and then the slides incubated O/N for final reading. Because the reaction is forced, it may in some instances increase the sensitivity as well as it decreases the time from 24-72 hours to less than 24 hours. Variables include the type and charge on the agar/agarose used in the gel as well as the pH of the electrophoresis buffer. The charge on the agar/agarose can be used to adjust the location of the bands.

This method has been used to titer antibodies to Sm and RNP. Antibodies to RNP are present in high titer and are the only ENA antibodies in MCTD. Serial dilutions of the patient's sera are tested in parallel. One set of dilutions is tested using untreated thymus extract as the antigen while the parallel set is tested using thymus extract heated at 56° C for 30 minutes and cooled rapidly. RNP reactivity is destroyed by heat treatment while Sm reactivity is not. If the serum contains antibodies to RNP alone, there will be a reaction only with the untreated extract. If the serum contains antibodies to both RNP and Sm, the highest titer observed with the untreated extract is the titer for anti-RNP and the titer in the heated extract is the titer for anti-Sm. If the titers are equal, the reactivity is to Sm only.

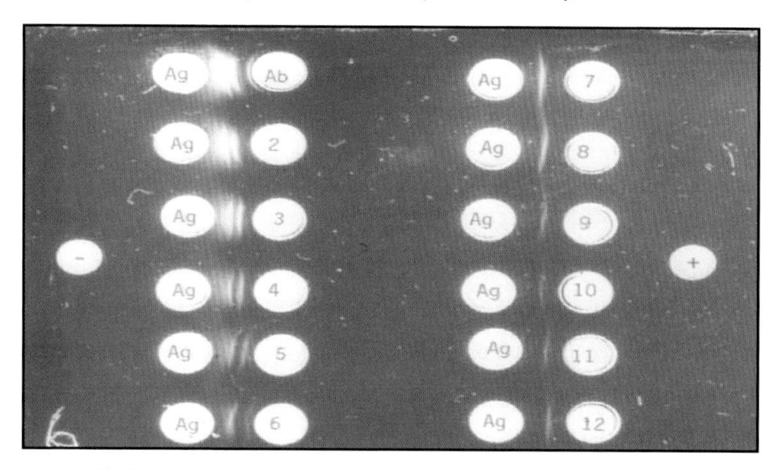

Figure 1. CIE Slide. This CIE slide demonstrates the titer of a patient containing antibodies to both Sm and RNP. The antigen is rabbit thymus extract. Well #1 contains the undiluted patient serum and wells #2-12 contain doubling dilutions of the serum ranging from 1:2-1:2048. The titer of the Sm antibodies is well #6 or 1:32. The titer of the RNP antibodies is >1:2048.

1. Kurata N and EM Tan. Identification of antibodies to nuclear acidic antigens by counterimmunoelectrophoresis. *Arthritis Rheum.* 19: 574-580. 1976.

2. Nakamura RM, CL Peebles, RL Rubin, DP Molden and EM Tan. Autoantibodies to nuclear antigens (ANA). Counterimmunoelectrophoresis. In: *Advances in laboratory tests and significance in systemic rheumatic diseases, 2nd ed.* ASCP Press, Chicago, Ill. 1984. pp 116-121.

Hemagglutination

Hemagglutination assays use antigen coated on red blood cells (RBC) to test for the presence of certain antibodies by agglutination. These assays are very sensitive. Agglutination assays require a lesser amount of specific antibody to demonstrate a reaction than do precipitation assays. IgM antibodies agglutinate better than IgG antibodies. Control and patient serum dilutions are prepared in microtiter plates. The antigen-coated RBCs are added to each of the wells, mixed and allowed to settle. Reactions are determined by the mat pattern on the bottom of the well. This procedure was the early method for quantitation of Sm and RNP [1]. In the procedure cells are first coated with an extract of rabbit thymus acetone powder (RTE). A portion of these cells is then treated with RNase to remove the RNP reactivity. Sera from patients and controls are titered in two parallel rows of microtiter plates. The untreated and the treated cells are then added. The plates are set aside to allow the cells to settle. A test is considered positive if there is a smooth mat of cells covering the bottom of the well and negative when there is a clearly defined button. Intermediate results are considered negative. If there is no agglutination in either of the rows, the test is considered negative. If agglutination is positive in the untreated cells and negative in the RNase treated cells, the results indicated the presence of antibodies to RNP only. If agglutination occurs in the untreated cells and also in the RNase treated cells, the serum contains antibodies to both Sm and RNP. The titer of the RNP is the titer observed in the untreated cells and the titer of the Sm is the titer in the RNase treated cells. The presence of high titered antibodies to RNP alone in the setting of an overlap syndrome is diagnostic for MCTD.

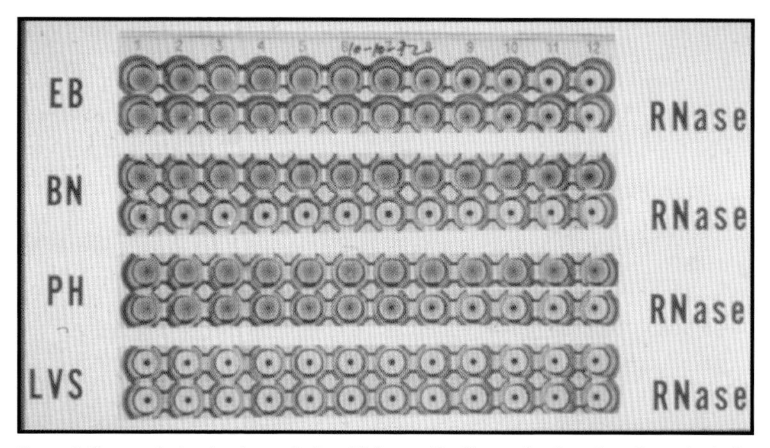

Figure 1. Hemagglutination Assay. Patient EB has antibodies predominately to Sm as both the treated and untreated rows are the same. Patient BN has antibodies to RNP only. Patient PH has a high titer of antibodies to RNP and a low titer of antibodies to Sm. Patient LVS is negative. [2]

1. Nakamura RM, CL Peebles, and EM Tan. Microhemagglutination test for detection of antibodies to nuclear Sm and Ribonucleoprotein antigens in systemic lupus erythematosus and related diseases. *Amer. J. Clin. Path.* 70: 800-807. 1978.

2. Tan EM and C Peebles. Quantitation of antibodies to Sm antigen and nuclear ribonucleoprotein by hemagglutination. In: Manual of Clinical Immunology 2nd ed. NR Rose and H Friedman, ed. ASM Press, Washington, DC. 1980. pp 866-870.

Immunoblot Assay

Immunoblot assays are often termed "western blot assays".

In immunoblot assays, antigens are adsorbed to a membrane. The membrane is incubated with the test sera, then the conjugate and finally any bound conjugate is detected. In the standard method a cell extract is separated by SDS- polyacrylamide gel electrophoresis (SDS-PAGE) resulting in a series of protein bands separated by size [1]. In this method a soluble suspension of cellular protein is prepared with Laemmli sample buffer [2] prior to applying it to the gel. The proteins become denatured during preparation. The separated proteins in the gel are then electrotransferred to nitrocellulose. During this time, with the removal of the SDS, some of the proteins become renatured partially or completely. Vertical strips of the nitrocellulose are cut and probed with controls and patients sera. Antibodies to some conformational epitopes are detected, but epitopes formed by protein-protein interactions are not detected. The test proceeds as in any other indirect method. The detecting reagent varies depending on the laboratory and kit. In some instances, the probe may be an enzyme-labeled antibody producing a colorimetric result. In other instances, the probe may have a chemiluminescent tag and be processed using photographic film. In the figure below the probe used was ^{125}I–Protein-A. The Protein-A binds to most classes of IgG. If there are IgG antibodies reacting with any proteins on the nitrocellulose, bands will be detected. Antibodies to many known antigens give characteristic banding patterns. Care must be taken in the interpretation of individual protein bands as there are many proteins present in the extract that migrate in similar locations [3].

In a variation of this assay, specific antigen is applied directly to nitrocellulose. In some instances this may be the application of a single specific antigen (dotblot) or multiple specific antigens applied in a banded pattern. This technique, unlike the standard immunoblot, allows for the detection of only those antigens of specific interest. Like ELISA, the purity of the antigen is very important.

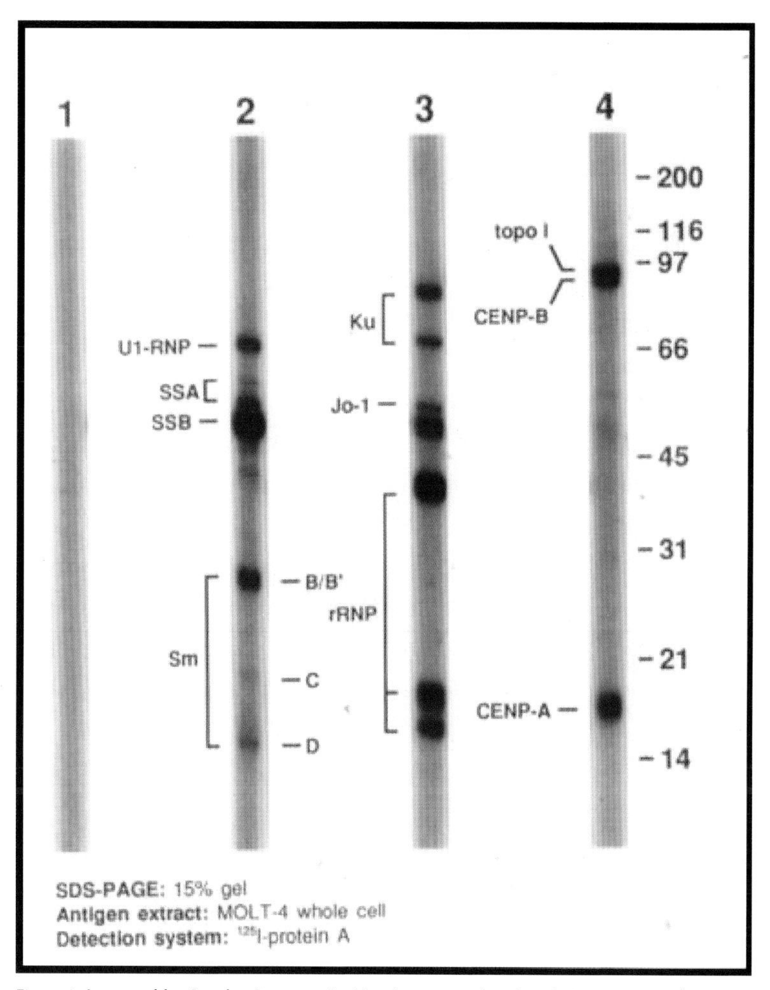

Figure 1. Immunoblot Results. A composite blot demonstrating the relative position of proteins associated with multiple ENAs. The brackets link proteins associated with a specific antibody reactivity.

1. Takács B. Electrophoresis of Proteins in Polyacrylamide Slab Gels. In: Immunological Methods, Academic Press, Inc. 1979. pp 81-105.

2. Laemmli UK. Cleavage of structural protein during the assembly of the head of bacteriophage T4. *Nature (London)* 227: 680-685. 1970.

3. Chan EKL and KM Pollard. Detection of Autoantibodies to Ribonucleoprotein Particles by Immunoblotting. In: Manual of Clinical Laboratory Immunology, 5th ed. Rose NR, EC de Macario, JD Folds, HC Lane, and RM Nakamura ed. ASM Press, Washington, DC. 1997. pp 928-934.

Immunoprecipitation

Immunoprecipitation for the identification of the extractable nuclear antigens involves the reaction of an antigen source, most often a cell extract, with antibodies present in a dilution of the serum to be tested. The reaction occurs in solution. The antigen-antibody complex formed is captured on a bead, the antigen and antibody dissociated, and the components separated on a gel to identify the antigen present in the complex. The procedure is used to identify both the proteins that are bound and/or the nucleic acids bound to them. This procedure is more often performed in research laboratories as radiolabeled extracts are commonly used. In general, an extract is prepared from cells that have been radiolabeled with either ^{35}S-methionine or ^{32}P-phosphate. The antigens are presented in a more native conformation so antibodies to conformational antigens are detected. A dilution of patient or control serum is then mixed with a small amount of the labeled extract along with protein-A-sepharose beads and allowed to react. Protein-A is produced by *Staphylococcus aureus*. It has the property to bind to IgG, especially aggregated IgG as in the complexes [1]. By coating sepharose beads with the Protein-A, the binding produces aggregates easily separated by centrifugation. If the patient has antibodies that react with proteins in the extract, an antigen-antibody complex will be formed. This complex then binds to the protein-A-sepharose bead to be separated by centrifugation. The bound complexes are washed well to remove any unbound antigen and the complex dissociated. The radiolabeled dissociated proteins are then separated by SDS-PAGE. The resulting gel is dried on filter paper. The dried gel is put into a film cassette and a sheet of photographic film added. The cassette is placed at -70^{0} C for exposure. The exposed film can be processed in any standard photographic processor. The ENAs precipitate specific proteins or complex of proteins. If the specific protein is part of a complex, the whole complex will be precipitated. For example, the Sm/RNP complex is precipitated by antibodies to both Sm and RNP. Care must be taken in interpretation of the reactions as multiple proteins may migrate at the same size. [2]

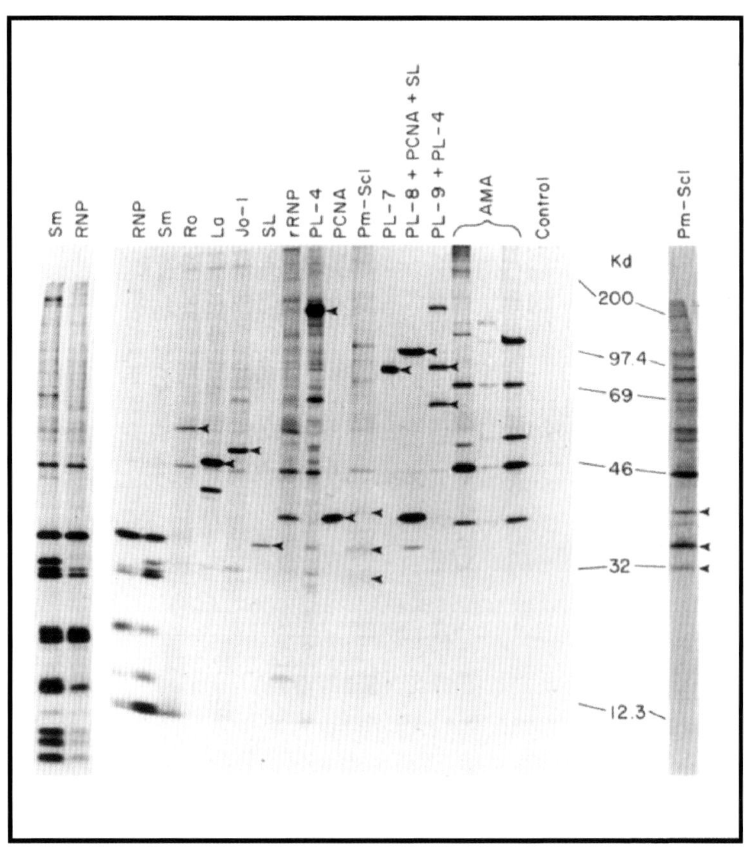

Figure 1. Immunoprecipitation patterns of various ENAs.

1. Kessler SW. Rapid isolation of antigens from cells with a staphylococcal protein A-antibody adsorbent: parameters of the interaction of antibody-antigen complexes with protein A. *J. Immunol.* 115: 1617-1624. 1975.

2. Tan EM and CL Peebles. Immunoprecipitation of labeled proteins. In: Manual of biological markers of disease. Van Venrooij WJ and RN Maini, ed. Kluwer Academic Publishers, Dordrecht, The Netherlands. 1996. pp A9.1--A9.13.

Crithidia luciliae

Antibodies to nDNA are often detected using IIF slides with *Crithidia luciliae*, a hemoflagellate as substrate. *Crithidia luciliae* have a large kinetoplast that contains circular nDNA without any single stranded regions or histones. This allows the detection of antibodies to nDNA and not to other nuclear antigens. The procedure is a standard IIF procedure in which the serum dilutions of the patient and the controls are added to individual wells on the slide. The normal screening dilution is 1:10. The samples are incubated for a standard amount of time, usually 30 minutes, washed and then incubated with conjugate. The current recommendation is for use of IgG specific conjugates. Low levels of IgM antibodies to DNA may be detected in conditions other than SLE. This assay does not include the addition of ammonium sulfate so it detects a broader spectrum of nDNA antibodies than the Farr assay.

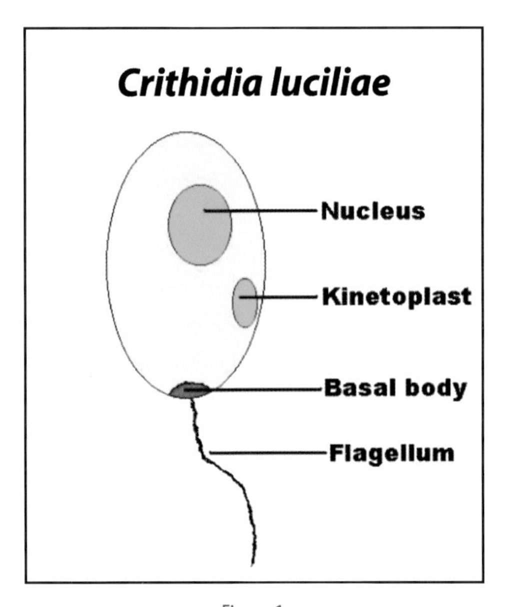

Figure 1.

1. Aarden LA, ER deGroot, TEW Feltkamp. Immunology of DNA. III. *Crithidia luciliae*, a simple substrate for the determination of anti-dsDNA with the immunofluorescence technique. *Ann. N.Y. Acad. Sci.* 254: 505-515. 1975.

2. Nakamura RM, CL Peebles, RL Rubin, DP Molden, and EM Tan. Native-DNA autoantibodies: indirect immunofluorescence assay with *Crithidia luciliae.* In: *Autoantibodies to nuclear antigens (ANA). Advances in laboratory tests and significance in systemic rheumatic diseases, 2nd ed.* ASCP Press, Chicago, Ill. 1984. pp 82-87.

Farr Assay

The Farr assay was the earliest method for the detection of antibodies to nDNA [1]. The test serum is mixed with a defined amount of ^{14}C-labeled DNA and allowed to react. This results in a combination of free ^{14}C-DNA and ^{14}C-DNA+antibody complexes. Ammonium sulfate is added to the mixture to form a 50% solution. The addition of the ammonium sulfate results in the precipitation of the ^{14}C-DNA+antibody complexes while the uncomplexed ^{14}C-DNA remains in solution. The

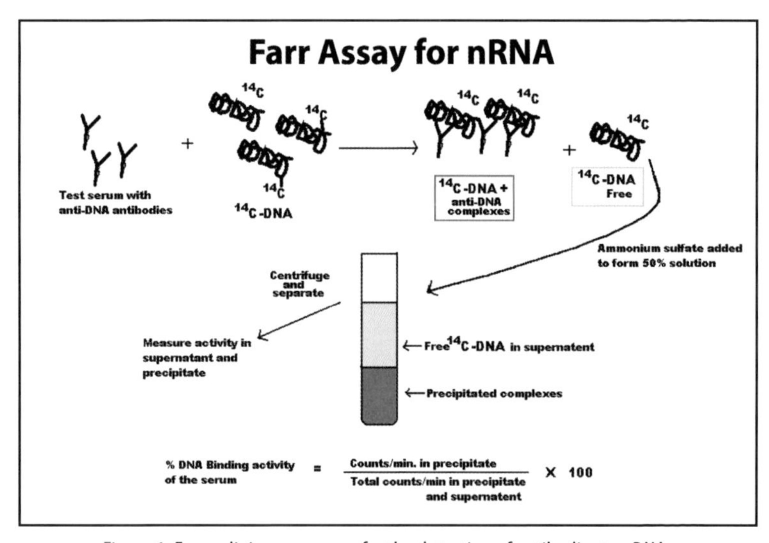

Figure 1. Farr radioimmunoassay for the detection of antibodies to nDNA.

samples are centrifuged to separate the free and bound ^{14}C-DNA. The use of ammonium sulfate results in the detection of antibodies that bind nDNA in high ionic strength. The % DNA binding activity of the serum is calculated by the counts/minute in the precipitate divided by the total counts/minute in the precipitate plus the supernatant multiplied by 100. High binding in this assay by SLE patients is historically associated with renal involvement. Occasional false positives may occur due to the binding of certain serum components other than immunoglobulin to DNA. These complexes may also be precipitated by ammonium sulfate. This assay has largely been replaced by the *Crithidia luciliae* IIF and ELISA.

1. Wold RT, FE Young, EM Tan, and RS Farr. Deoxyribonucleic acid antibody: a method to detect its primary interaction with deoxyribonucleic acid. *Science:* 806-807. 1968.

ELISA

ELISA has become the method of choice for the detection of many of the major antibodies associated with SRD. These include nDNA, chromatin, Sm, RNP, SS-A/Ro, SS-B/La, Scl-70 and Jo-1. The standard assay is an indirect solid phase assay with the antigen coated onto the wells of a 96 well microtiter plate. A specific amount of the control and patient serum dilutions are added to duplicate wells in the plate. Controls should include a blank, a low positive control, a high positive serum, and a normal human serum. The reaction proceeds for a specified amount of time and if it is present, any reactive autoantibody binds to the antigen on the well. The plates are washed well and the enzyme labeled secondary antibody is added. The plates are incubated for a specified amount of time and the enzyme labeled anti-human antibody binds to any bound patient antibody. The plates are washed well. The chromogen substrate for the reaction is added and the plates incubated for a specified amount of time. The presence of bound enzyme catalyzes a reaction in the substrate resulting in color formation. This reaction time is critical. The time is set so that the reaction is read on the slope of the curve and not on the plateau. A reagent is then added to stop the reaction. The plates are read in a spectrophotometer and the O.D. recorded.

Most commercial assays are reported in units generated from a standard curve. The absorbance in the reagent only well should by <0.1 OD. If this well is too high, it could mean that the plates were insufficiently washed. Increase the number of washes. If the NHS is >0.2 it may indicate incomplete blocking of the wells. If the result is not consistent with the clinical history, rule out a heterophile antibody. With the use of monoclonal antibody treatment in cancer, the number of patients with antibodies to mouse proteins has increased [1].

While ELISAs from the majority of manufacturers will detect most antibodies to the ENAs quite well, they tend to vary somewhat in their detection of antibodies to Sm. The most variation occurred in the detection of antibodies to nDNA [2]. The major reasons for these variations are the type and source of the antigen used and the specificity of the conjugate.

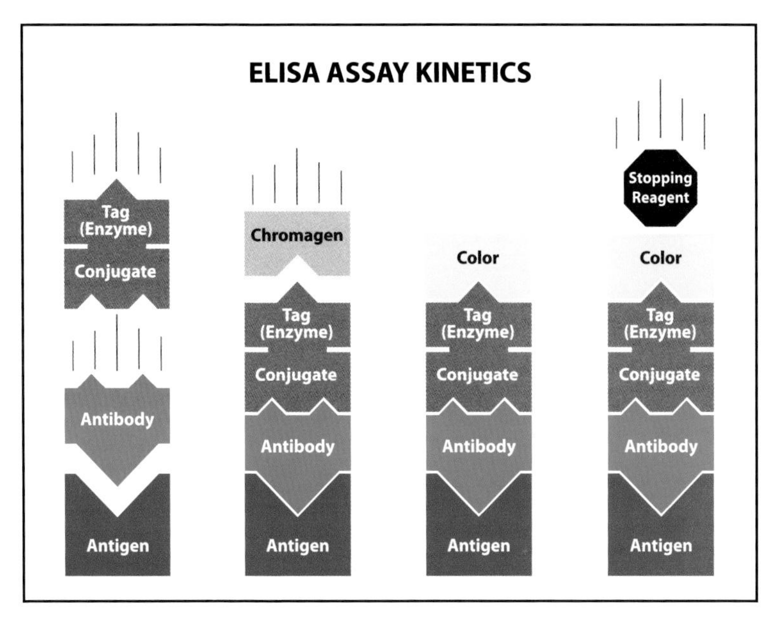

Figure 1. Diagram illustrating the assay kinetics for ELISA. (Courtesy of Doug Kibbey, INOVA Diagnostics, Inc.)

Antigens bind to the plate by a combination of charge, van der Waals and hydrophobic interactions. Each antigen is unique in its binding ability. The type of antigen used to coat the plates is one of the most important variables. Some antigens may come from animal tissue while others may be recombinant proteins. They may be purified by any number of different techniques. Both the source of the antigen and the method of purification will affect the antigenic reactivity. The conditions for the optimum coating procedure vary for each antigen. Variables include the type of coating buffer, the concentration of the antigen, and the surface properties of the microtiter plates. Most antigens are coated in a range of 1-50 μg/ml. nDNA will not bind directly to microtiter plates so a linker molecule is added to which the nDNA can bind. The choice of linker will affect the results of that assay.

Not only the specificity of the detecting antibody but the amount of enzyme label per specific antibody will affect the results of the assay. Some detecting reagents are anti-IgG specific while others will detect not only IgG but IgA and IgM as well. The popular enzymes for ELISA are alkaline phosphatase and horseradish peroxidase.

The advantages of ELISA are that it is relatively inexpensive, is adaptable to automation, produces objective results, and gives better lab-to-lab reproducibility. But, when evaluating kits for use or determining the reasons for variability between assays both within and between laboratories, it is necessary to examine the nature of the antigens employed and the antibody detecting reagents in each kit.

1. Carpenter AB. Enzyme-linked immunoassays. In: Manual of Clinical Laboratory Immunology, 5th ed. Rose NR, EC de Macario, JD Folds, HC Lane, and RM Nakamura ed. ASM Press, Washington, DC. 1997

2. Tan EM, JS Smolen, JS McDougal, BT Butcher, D. Conn, R Dawkins, MJ Fritzler, T Gordon, JA Hardin, JR Kalden, RG Lahita, RN Maini, NF Rothfield, R Smeek, Y Takasaki, WJ van Venrooij, A Wiik, M Wilson, and JA Koziol. A critical evaluation of enzyme immunoassays for detection of antinuclear autoantibodies of defined specificities. I. Precision, Sensitivity, and Specificity. *Arthritis Rheum.* 42:455-464. 1999.